INDUSTRIAL ENGINEERING

INDUSTRIAL ENGINEERING

N. J. MANEK

M.E. (Mech.)
Asst. Prof. of Prod. Engg.
B.V.M. (Engg. College)
Vallabh Vidyanagar
Gujarat

UNIVERSITY SCIENCE PRESS
(An Imprint of Laxmi Publications Pvt. Ltd.)
BANGALORE • CHENNAI • COCHIN • GUWAHATI • HYDERABAD
JALANDHAR • KOLKATA • LUCKNOW • MUMBAI • RANCHI
NEW DELHI • BOSTON, USA

Published by :
UNIVERSITY SCIENCE PRESS
(An Imprint of Laxmi Publications Pvt. Ltd.)
113, Golden House, Daryaganj,
New Delhi-110002

Phone : 011-43 53 25 00
Fax : 011-43 53 25 28

www.laxmipublications.com
info@laxmipublications.com

Price : **Rs. 90.00** *Only.*

New Edition : 2001
Reprint : 2008

OFFICES

India

✆ **Bangalore**	080-26 61 15 61
✆ **Chennai**	044-24 34 47 26
✆ **Cochin**	0484-239 70 04
✆ **Guwahati**	0361-254 36 69, 251 38 81
✆ **Hyderabad**	040-24 65 23 33
✆ **Jalandhar**	0181-222 12 72
✆ **Kolkata**	033-22 27 37 73, 22 27 52 47
✆ **Lucknow**	0522-220 95 78
✆ **Mumbai**	022-24 91 54 15, 24 92 78 69
✆ **Ranchi**	0651-221 47 64

USA

Boston
11, Leavitt Street, Hingham,
MA 02043, USA

UIE-9236-090-INDUSTRIAL ENGINEERING

C—15581/08/03

Typeset at : Goswami Printers, Delhi.

Printed at : Sudha Offset Press, Delhi.

DEDICATED
TO
MY FATHER & MOTHER

Preface

Industrial Engineering has acquired great importance in the fast developing technology today. Technology has brought about profound changes in methods of operation, whether at the factory or in the office. There has been a continued growth in the services sector, mostly at the expenses of the Industrial and Agricultural sectors. Simultaneously, new and different arrangements of working were introduced in a multitude of enterprise.

Most of the universities, polytechnics and IITS today offer course on Industrial Engineering at either the under graduate level or the diploma level. This course covers, those techniques, particularly method study, work measurement, Ergonomics, SQC which are used to increase the productivity of the country and standard of living of the people.

Industrial Engineering is not a isolated course from other courses. It is a management course. It requires the knowledge of manufacturing processes, engineering metrology, materials to select the method of production, materials, etc.

In early days, people were ready to pay higher for quality. Now a days trend has been changed. People want good quality at a lower price. There is a competition among the producers to produce quality goods at lower cost. Industrial Engineering helps in reducing the price. Industrial Engineering is a valuable "tool" of management. Implementation of Industrial Engineering concept involves little or no capital expenditure on facilities and equipments.

I am thankful to my colleagues shri T.B. Pankhania and Shri B.C. Shah of Mechanical Engineering department for encouragement and valuable discussions. I will fail in my duty if I don't mention the tolerance and encouragement of my wife Jyoti in the present endeavour. I am also thankful to Saumya Gupta, Director of Laxmi Publications Pvt. Ltd. for encouragement to write this book.

—N.J. Manek

Contents

1

Industrial Engineering and Scientific Management

1.1. INTRODUCTION

Industrial Engineering is concerned with the design, improvement and installation of integrated system of people, materials and equipment. It draws upon the specialised knowledge and skill in mathematical, physical and social science together with the principles and methods of engineering analysis to predict and evaluate the results to be obtained from such systems.

Industrial Engineering is unique in the sense that it is a link between technology and management, technology and economics and technology and science. Industrial Engineering discipline is different from other engineering disciplines. It can be stated that Industrial Engineering deals with people as well as things.

Industrial Engineering adopts wholistic approach in problem solving through application of scientific and practical knowledge. It acts like a surgeon's knife, lying bare the activities of a company and functioning, good or bad, for all to see. For this reason it must be handled, like the surgeon's knife, with skill and care.

1.2. APPLICATION OF INDUSTRIAL ENGINEERING

Industrial Engineering is engineering approach to the detailed analysis of the use and cost of the resources of an organization. The main resources are men, money, materials, equipment and machinery. The Industrial Engineer carries out such analysis in order to achieve the objectives (to increase productivity or profits, etc.) and policies of the organization. The Industrial Engineer is engaged in the design of a system and his function is primarily that of management. Industrial Engineering can be with manufacturing applications, *e.g.* Method study, Work measurement, Incentives, and plant layout. The lingering tendency to consider Industrial Engineers as 'Stop Watch Engineers' is no longer appropriate.

Industrial Engineering has been a challenging one. In the early days, Industrial Engineering was largely time study oriented. By the years passed, the dynamic nature of modern Industrial Engineering to serve all levels of managements in problem solving are applicable in almost every kind of organization *viz.* banks, hospitals, government at all levels, transportation, construction, ware-housing, manufacturing etc. Thus, Industrial Engineering covers a broad cross section of business, industry and institution. It is a valuable "tool" of management. It is a means of raising the productivity of a plant by the reorganization of work, a method which normally involves little or no capital expenditure on facilities and equipments. It is a "tool" which can be applied everywhere. It can be used with success wherever work is done or plant is operated, not only in production shops but also in offices, stores, laboratories and service industries such as wholesale and retail distribution centre and restaurants and on farms. It is one of the most penetrating tools of investigation available to management. It involves investigation by direct observation of all the factors affecting the efficiency of a given operation.

1.3. NATURE OF INDUSTRIAL ENGINEERING

Industrial Engineering is a service function. It must always be the responsibility of someone who is able to undertake it full time, without direct management duties ; someone in a 'staff' and not a 'line' position. People in a 'staff' position means strictly advisors with no power or authority to put their recommendations into operation. There function is to provide expert information. Industrial Engineering is a service to management and supervision.

Industrial Engineer gathers facts concerning the problems, subjects them to analysis, forms tentative conclusions, compares and tests various alternatives and finally presents his findings and recommendations to top management. For doing this he uses the tools and techniques of Industrial Engineering.

1.4. SCIENTIFIC MANAGEMENT TECHNIQUES

F.W. Taylor was the founder of scientific management techniques. In his opinion, management function consisted of three major duties ;

(*i*) application of scientific approach to every work in place,

(*ii*) selection of the right person for a task and if necessary to train him for further improvement, and

(*iii*) to develop a spirit of co-operation with the workers.

Scientific management is the application of the scientific method to the study of the operations of large, complex organizations or activities. It encompasses the development of models for specific problems. The theory of scientific management is based on observation, measurement, experimentation and logical inferences therefrom results in maximum prosperity for both, the employer and the employee.

The fundamental objective of Industrial Engineering is to optimise the use of human resources and material resources available with an organization for economical production. This can be achieved by (*i*) the most effective use of the plant, equipment and human efforts, and (*ii*) evaluation of human work and selection and training of personnel. The former is known as work simplification and the latter as work measurement.

When work study emerged in the early part of this century as a technique aimed at rationalizing and measuring work, the emphasis was on economy of motion and movement. Because of that it was called time and motion study. Later on, it began to encompass other aspects of observing and analysing work, simultaneously, in the late 1940s and later on in the 1960s, other disciplines were developed, namely Industrial Engineering and production management respectively. These disciplines differed from work study in the sense that they were devoted to increasing the efficiency of a production operation as a whole, not just the methods of work. Thus modern production management looks at various aspects of production such as product design, quality control, plant layout and materials handling, production, planning and control, maintenance management and invariably work study. Advances in the field of computers and information systems helped to boost production management techniques to the level attained at present.

2

Production, Productivity and the Standard of Living

2.1. THE STANDARD OF LIVING

The primary responsibility of any National Government is to provide a decent standard of living to its countrymen. The standard of living means the degree of material well-being available to a person or class or community which is necessary for sustaining and enjoying life. The standard of living depends upon the prosperity of the nation which in turn depends to a considerable extent on its industrial advancement and economic growth. This industrial advancement and economic growth of the country again influence by natural resources, human resources, and technological advancement. This is the reason that the standard of living of the person or family in the different countries of the world varies greatly from country to country and even, within each country, from community to community and area to area. The standard of living of the person or family in the advance countries like U.S.A., Canada, Germany is higher compare to person or family in the developing countries like India.

There is a great influence of natural resources such as fertility of land, forest wealth, minerals, water resources etc. Though, the presence of these resources is not enough for economic development of a country. The proper exploitation of the resources and utilization through improved techniques are of prime importance. It reduces waste and makes possible their utilization for a longer time.

Human resources are another important factors in the economic development of the country. The general level of education is an important factor in national productivity. The use of computers and other sophisticated equipment and systems requires better educated employee. The level of education, knowledge, technical know-how and efficiency are low in developing countries like India. The educated and trained labour force leads to rapid economic growth. Government can help by sponsoring more education. Employees should be motivated to be productive. Behavioural scientists believe that individuals are motivated to act in a certain way by a desire to satisfy certain needs. Abraham Maslow postulated that there are certain essential needs for every individual and that these needs arrange themselves in a hierarchical pattern. Maslow argues that it is only when one need becomes largely satisfied that the next need in the hierarchy will start to exert its motivating influence.

At the bottom of the hierarchy are physiological needs. The next needs in the hierarchy are security, affiliation, recognition and fulfilment.

In practice, most people satisfy some of these needs in part and are left with some that are unsatisfied. In developing countries people are probably preoccupied more with their basic needs. In developed countries, physiological and security needs are normally met in large part, people would seem to be motivated more by needs, at the upper end of the hierarchy.

The third factor influence the economic development of the country is the technological progress which helps in raising output at reduced cost, thus improving the competitive potential in the market. Technological progress may be possible through research and development which may lead to many advantages which include (*i*) discovery of unknown resources, (*ii*) economical extraction of raw materials and (*iii*) decrease in processing cost of raw materials. Government can help by (*i*) promoting R & D in Industries and Universities, (*ii*) encouraging person saving and reducing taxes on profits so that people invest in new facilities, (*iii*) allowing depreciation rates that will provide cash flow for new investment, and (*iv*) directly encouraging new investment through increased investment tax credits.

2.2. REQUIREMENTS FOR A MINIMUM SATISFACTORY STANDARD OF LIVING

In too many parts of the world the ordinary man is still hardly able to satisfy his basic needs. By the year 2000 close to one billion people will be living below the poverty line and struggling to meet their basic needs. The basic needs must be met in order to attain a minimum decent standard of living.

These basic needs are : (1) Food (2) Clothing (3) Shelter (4) Security and (5) Health and essential services. The first three things which a man has to obtain for himself. People must pay for it either in the form of money or work. The food is required to generate the energy needed for living and working. More and cheaper food is obtained by increasing the productivity of agriculture. By using better seed, better methods of cultivation and more fertiliser, the productivity of agricultural land can be raised.

Clothing is required to protect the human body from environment. It also permit bodily cleanliness. Shelter provides protection to the household equipment and furniture.

Security and essential services are generally provided by national government and other public authorities. Security against violence, unemployment, personal sickness or old age provided by government.

People should get safe drinking water, sanitation, medical care, education and a means of transport. More and cheaper clothing and shelter are obtained by increasing the productivity of Industry. More security and essential services are obtained by increasing all productivity and earning power.

The greater the amount of goods and services produced in any country or community, the higher its average standard of living will be. For a society or a nation to raise the standard of living of its people, it must strive to maximize the return from its resources or improve productivity so that the economy can grow.

2.3. PRODUCTION

With reference to goods, the production is defined as a fabrication of a physical object through the use of labour, materials, and equipment ; whereas, with reference to services, the production is the discharge of a function which has some utility. This definition of a production activity is a very broad one. The essence of production is the creation of goods, may be by the transformation of raw material or by assembling so many small parts (as in T.V. or watch manufacturing). Production in every day life can be noticed in factories, hospitals, offices, banks etc. To improve the standard of living, it is necessary to increase the production. Production can be increased by increased employment, duplication of machines, increasing the number of shifts or by installation of highly productive machines. In this case, increased production (output) require increase in input resources. This may or may not result in increased productivity. The another way to achieve higher production is to raise productivity ; that is to get maximum output of the available resources. The resources may be land, materials, equipments, services of men or any combination of these factors.

Production may be measured in terms of output per man-hour or machine-hour. In manufacturing concern, output can also be expressed in numbers, by value and by conformity to predetermined quality standards. In a service enterprise such as a public transport company or a travel agency, the output is expressed in terms of the services rendered. In a transport company this may be the number of customers or tonnes of cargo per kilometer carried. In a travel agency it could be value of tickets sold or average value of tickets per customer.

2.4. PRODUCTIVITY

The productivity may be defined as the ratio of output and input. Thus,

$$\text{Productivity} = \frac{\text{Output}}{\text{Input}} \quad ...(2.1)$$

The efficiency of the machine is also defined as the ratio of output and input. Thus productivity of a production system is analogus to the efficiency of a machine. Machine efficiency generally less than 100%, but productivity some time may be 100% . The definition of productivity applies in an enterprise, an industry or an economy as a whole. The term "productivity" can be used to measure the extent to which a certain output can be extracted from a given input. Output means the amount produced or the number of items produced. The output can also be measured in other units such as output per unit of floor areas, output per unit of energy consumed, output per unit capital etc. The input means various resources employed, in the course of production. These resources may be land and building, materials, plant, machines and tools, and services of men or, as is generally the case, a combination of all the above resources. From the definition of the productivity, we can say that it is nothing more than the arithmetical ratio between the amount produced and the amount of resources used. It can be seen from expression (2.1) that the productivity can be increased by three different ways :

(*i*) Maintaining input and increasing the output,

(*ii*) Maintaining output and decreasing the input, and

(*iii*) Increasing small amount of input to get larger amount of output.

In all the above cases, the attempt is made to increase the ratio of output and input. This means that the production is increased. But an increase in production does not by itself indicate an increase in productivity. If the input of resources goes up in direct proportion to the increase in output, the productivity will stay the same. If input increases by a greater percentage than output, higher production will be being achieved at the expense of a reduction in productivity.

One may speak of productivity of labour, capital, material, power, machine, etc. or any combination of the inputs. In practice the choice of factors is limited to those cases where measurement of productivity are possible. It appears simple enough to measure the productivity in cases where both the output and the input are tangible and can be easily measured, but it is more difficult to estimate productivity once intangibles are introduced. The factors affecting productivity in an organization are many, and often interrelated. Productivity may be used to measure increase on output expressed in terms of quantity produced or in monetary terms. These are tangible factors. The intangible factors may be quality of work, users satisfaction, etc. The practical meaning of productivity is the effective utilization of resources or "input factors". Here the available resources are to be used effectively to increase the productivity. There is no absolute value for productivity in any given situation and when one talks of higher productivity, it is relative to the past period, or to some other standards, which may be another similar production unit or other nation.

Measurement of productivity provides information as to the level attained, rate of growth, utilization of resources and consequently pin points the area and factors which impede productivity at various levels. It is thus comparative assessment at various levels. The overall productivity of industry or any other factors, would be of great interest in the general economic analysis.

Productivity also influence the standard of living. Higher productivity can contribute to a higher standard of living. Higher the productivity means more production at same cost, or same production at less cost, there is a gain to the community as a whole. The members of community acquires more and better goods and services and improve their standard of living.

2.5. FACTORS INFLUENCING PRODUCTIVITY

Productivity is implied in every economic activity and higher productivity can be regarded as efficient use of the inputs. Productivity may be influenced by a series of external factors, as well as by a number of internal factors. External factors include the availability of raw materials and skilled labour, government policies towards taxation, existing infrastructure, capital availability and interest rates. These external factors are beyond the control of any one organizer. Industrial Engineering also can not help to raise the productivity of these external factors. Other factors, however within the control of employers and that will be discussed here. The other factors are the inputs or resources in terms of land and building, materials, machines and equipments, energy, and human services.

Land and Building

Location of land and building plays an important role to raise the productivity. Industries should be located where labour costs are low compared with material costs, or compared with the capital invested in plant and equipments. Effective utilization of plant and building also contribute to raise the productivity of these resources. The productivity of plant and building is increased by properly planning their layouts. The systematic arrangement of machines, equipments and work-stations into the plant building reduce the unnecessary movements of men and materials. It will permit an uninterrupted flow of materials with minimum of handling in processing from raw material to finished product.

Materials

The cost of raw material influence the productivity. When the availability of the raw materials is poor, there is an important need for their economic utilization. Higher productivity can be achieved through proper use of materials. Effort must always be made to achieve optimum utilization of the materials available and to eliminate of their non productive use by reducing wastage or scrap. Sometimes a little change in the design of the component may save a lot of material. There are two types of materials. The direct materials that can be converted into products and indirect materials like oil, grease, cleaning materials. The productivity of materials can be increased by converting materials in to goods of higher value, substituting improved types or sizes of materials, reducing wastage of materials, producing useful bye-products, and reducing consumption of indirect materials.

Machines and Equipments

Machines, Tools and Materials handling equipments are required for manufacturing, handling and transportation of materials and finished goods. Productivity of the machines can be increased by using improved tools, simple attachments and other devices, cutting fluids, and appropriate speed, feeds and depth of cut. By reducing process time of machines by improving the process equipment design, eliminating unnecessary activities, rearranging or combining necessary process activities, productivity can be increased. Total production time can be cut short considerably by improving machine setting up methods, thereby reducing setup times. Proper maintenance of machines will reduce sudden break down and add to the productivity.

Energy

Energy includes its various forms such as electricity, gas, oil or solar power. Productivity of these indirect material can be increased by reducing their consumption and waste.

Human Services

The human factor is one of the most crucial factor in enterprise operations. To give the best of their ability, employees must be motivated to be productive. Pay is not enough ; they need to have good, safe, working conditions and to be recognized as the most vital part of the enterprise. Managers must be able to provide a motive to make people want to do it. Employees at all levels must feel a sense of belonging to the enterprise ; they should develop a sense of security, and the feeling that they are working in a safe, healthy and enriching working environment. When this happens they will contribute not only their labour but also many useful suggestions that can lead to productivity improvement. Productivity of man-power can also be improved by substitution of human power by mechanisation, reducing work content of a job by modifying product design, eliminating unnecessary activities, rearranging or combining necessary activities and improving material used, giving appropriate training to workers etc.

2.6. WORK CONTENTS

The amount of work involved in manufacture of a product is called work content. It is generally in time unit. It is time taken by a worker or a machine to carry out an operation or to produce a given quantity of a certain product. Work content is measured in "work-hours" or "machine hours". A work hour is defined as the labour of one person for one hour and machine hour means the running of a machine or piece of plant for one hour. The work content of a job is affected by many factors such as poor product design, improper selection of processes, short-comings on the part of management the attitude of human resources. The amount of total work content is divided into two parts : (*i*) Basic work content, (*ii*) Excess work content.

(*i*) **Basic Work Content.** It is the time taken to manufacture the product or to perform the operation if the design of the product or service provided is perfect, if the process or method of operation is perfectly carried out, the selection of process for its manufacture is most economical, the operator adequately trained to do the work and if there is no loss of working time from any cause whatsoever during the period of the operation. In short it can be defined as the irreducible minimum time theoretically required to produce one unit of output.

The situation mentioned above is perfect or ideal which never occurs in practice. A good proportion of excess time is added to the basic work content which is called excess work content. Fig. 2.1 shows that how the total time of a job is made up.

(*ii*) **Excess Work Content.** As mentioned above that ideal situation never occurs in practice. There is always some extra work is added to the basic work content. These extra work may be due to poor design, material selection, inefficient work method, poor layout, ineffective process selection or contribution of human resources. It is not possible to eliminate this excess work content completely. Therefore, the task of management is to apply corrective measures in order to reduce excess work content to a minimum.

(A) *Work content added by poor design of product or improper utilization of materials.*

(*i*) **Poor Design.** There are a number of factors to be considered before a design is prepared, finalised and issued to shop. The product may require a large number of non-standard parts causing a lengthy time of assembly. Lack of standardization of product or their parts may add extra work. The parts produced in small batches, with time lost as the operator adjust and shifts from one batch to the next may add work.

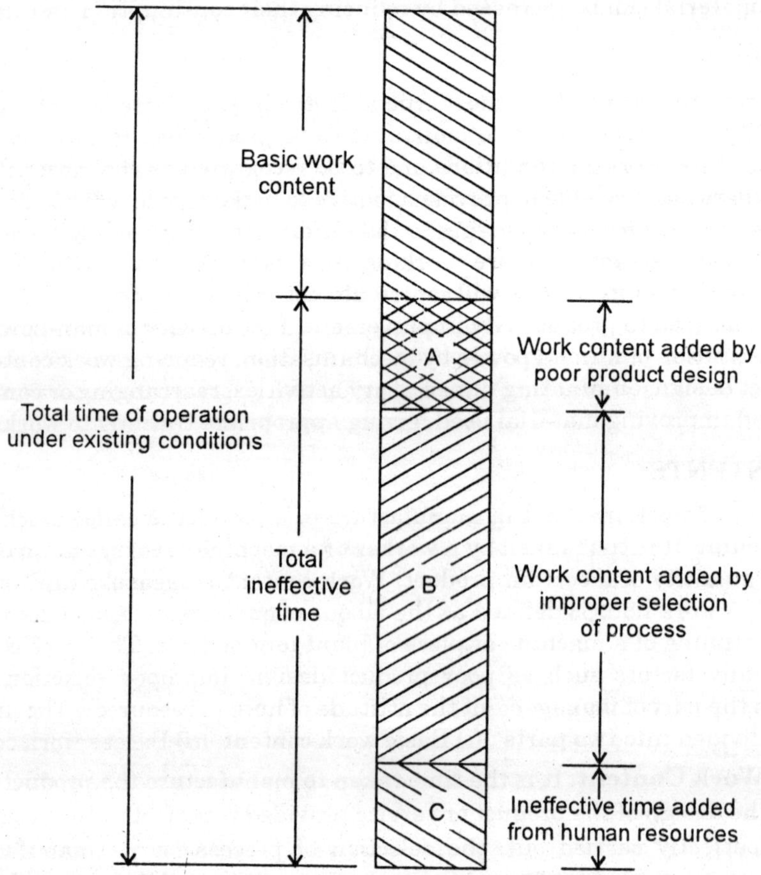

Fig. 2.1. Basic and excess work content.

(*ii*) **Manufacture of Irregular Shaped** component by machining instead of by die-casting or other suitable process. Thus an excess amount of material has to be removed to bring them to their final shape. This increases the work content and wastes material as well. Cutting operation in particular need require careful examination to see if the resulting waste can be reduced to a minimum.

(*iii*) **In Correct Quality Standard** may increase work content. It is necessary to be critical about tolerances and the motto should be to give as liberal tolerances as permissible, consistent with the function of the component. Sometimes unnecessarily tight tolerances requires extra machining and a corresponding waste of materials. Too loose tolerances may result in a large number of rejects.

(B) *Work content added by poor selection of process or operation* :

(*i*) **Ineffective method of work.** A job can be fabricated in a number of different ways. The economics of manufacture should be considered in selection of any machine or process. By examining the way a certain operation is carried out and devising a better method, ineffective time can be reduced.

A part can be made on a centre lathe, a capstan lathe or a single spindle automate. Here quantity is involved in selection of machine. For a few pieces, a centre lathe is economical. Automate is economical for the largest number of pieces. The actual time depends upon the setting time and machining time of the job on each machine. The selection of right cutting tools, use of correct cutting speeds, feeds and depths of cuts are also important factors influence in reduction of production time.

(*ii*) **Poor plant layout and materials handling equipments.** Proper plant layout and effective use of space reduces wasted movement of man-materials, time and effort. Proper layout also influence the selection of materials handling system. Proper layout reduces the movement of material, parts and finished product in a plant. This will reduce the materials handling equipments. The use of appropriate materials handling equipment can save time and effort.

(*iii*) **Poor planning.** The proper planning and control of production operations can reduce production cost by increasing batch size. To achieve this the varieties of product should be restricted to a minimum. This reduce frequent stoppages as production changes from one batch to another.

(*iv*) **Poor planning of inventory.** The flow of material should be well planned so that there is no hold up due to lack of materials at any stage. The raw material is usually ordered and stocked ahead of time. The various inventories such as raw material, semi finish product, and finished products represents a tied-up investment. A proper inventory control can minimize such an idle investment and ensures that the operator do not run out of the material needed.

(*v*) **Poor maintenance of machines and equipments.** Poor maintenance results in frequent breakdown of machines and equipments. It is a loss of time until machines or equipments gets repaired. Preventive maintenance system would ensure the smooth functioning of machines and equipments. It reduces the frequency of breakdowns. Thus it minimizes excess work content of a job.

(C) *Work content added by human resources* :

(*i*) **Absenteeism and lateness.** A lot of time is wasted due to absenteeism, late coming, starting the work late and preparing for going early or deliberately working slowly. This is because of management fails to provide a safe and satisfying working environment. Suitable incentive scheme may curtail this waste.

(*ii*) **Accident hazards and safety.** All safety measures should be strictly followed. If safety precautions are not taken then accidents or occupational illness can occur, which effects on morale and increased absenteeism. Accidents can be reduced by providing good house keeping.

If these factors can be eliminated (an ideal condition which, of course, never occurs in practice), the minimum time and cost for the production of a given output and hence the maximum productivity is achieved.

3

Work Study and Human Factors

3.1. INTRODUCTION

Work study is a part of Industrial Engineering. Work study, as defined by British Standard Glossary of terms, is a "generic term for those techniques particularly 'Method study' and 'Work measurement' which are used in the examination of human work in all its contexts, and which systematically lead to the investigation of all factors which affect the efficiency and economy of the situation being reviewed, in order to effect improvement".

Work study is a effective tool of investigations in an organization. It is systematic both in the investigation of the problems and in the development of its solution. Work study finds the best and most efficient way of using available resources such as men, machine, materials and money. As the work study is systematic, it is time consuming. Therefore it is necessary to separate the job of making work studies from the other tasks.

Every organization tries to achieve higher quality production at lower minimum possible cost. This is possible by reducing the time of production and unnecessary activities during the production. This require to collect all the relevant data about the job or process. It is only by continuous observation and study at the work place or in the area where the activity is taking place that the facts can be obtained. The work study is the responsibility of some one who is in a staff and not a line position. The people in a staff position are strictly advisers with no power or authority to put their ideas into operations. The people in a line positions are directly supervisors with a power to order their staff of ranks below them. Work study is a service to management.

Work study is a valuable "tool" of management. There are many reasons summarized as follows :

* It is a means of raising productivity of a industry which normally involves little or no capital expenditure on facilities and equipment.
* As it is systematic, it covers every factors affecting the efficiency of an operation. No factor is over looked. It reveals all the facts about the operation, good or bad, for all to see.
* It is very accurate means of setting standard of performance.
* It improves safety and working conditions at work place by exposing hazardous operation.
* Work study is a universal tool which can be applied everywhere. It can be used in offices, stores, laboratories, and service industries.
* It is cheap and easy to apply.
* It is a penetrating tool of investigation. It attack on inefficiency in any organization. It involves investigation by direct observation of all the factors affecting the efficiency of an operation.

3.2. TECHNIQUES OF WORK STUDY

The fundamental objective of work study is to obtain the optimum use of human resources, materials and machines available with an enterprise for economical production. This can be achieved by effective utilization of plant and equipments, human efforts, selecting trained personnel, selecting appropriate materials and reducing wastage of materials. There are mainly two techniques of work study.

(*i*) Method study or work simplification.

(*ii*) Time study or work measurement.

Method study is defined as the systematic recording and critical examination of the activities in order to make improvements. Method study is concerned with the simplification of the job and develop more economical methods of doing it. Method study indicates how a job should be done to achieve higher productivity. It is concerned with the reduction of the work content of a job or operation.

Work measurement is defined as the application of techniques designed to establish the time for a qualified worker to carry out activities at a defined rate of working. It determines how long a job ought to take to achieve higher productivity. It is concerned with the investigation of ineffective time during manufacturing and with the subsequent establishment of time standards for the operation when carried out in the improved fashion, as determined by method study. Thus, method study and work measurement are closely linked. Both aims to achieve higher productivity. But method study should precede the use of work measurement.

3.3. BASIC STEPS OF WORK STUDY

As work study is systematic in investigation and problem solving, the correct procedure to be followed for effective results. It consists of eight steps in performing a complete work study.

(1) **SELECT** the job or process to be studied.

(2) **RECORD** all relevant information about the job or process by direct observations.

(3) **EXAMINE** the recorded information critically and challenge everything that is done.

(4) **DEVELOP** the most economical and practical method after careful consideration of all the alternatives.

(5) **EVALUATE** the quantity of work involved in the improved method and calculate a standard time for it.

(6) **DEFINE** the new method either verbally or in writing, using demonstrations.

(7) **INSTALL** the new method with the allotted time of operation.

(8) **MAINTAIN** the new standard method by monitoring and occasional checks.

Steps 1, 2 and 3 are common for both the techniques of work study. Step 4 is concerned with method study only. While step 5 is the part of work measurement only. Remaining steps 6, 7 and 8 occurs in both the technique.

In the application of the eight steps of work study described above, the human factors plays an important role. We shall now discuss the role of human factor in the application of work study.

3.4. WORK STUDY AND TOP MANAGEMENT

To establish a pharmaceutical store, it is required for the owner to be qualified in pharmacy degree. But there is no need of any qualification to establish any industry. Industrialist may get licence to establish an industry without any educational qualification or may be graduated in any discipline. Attitude of management is important towards the implementation of various scheme and suggestions made by work study department particularly when persons in top management are not engineering graduate. If the application of work study in an industry is to succeed, it is important to have the understanding and the backing of top management at all levels in an enterprise. If top management people such as the managing director, chairman or president of the company do not understand what the work study person is trying to do, it can not be expected that middle management and workers will lend their support either. It is the responsibility of the top management to make sure that the best use is made of all available resources. If the attitude of the management is indifferent towards suggestions made by the work study department nothing fruitful can be achieved. The enthusiasm and spirit of staff goes down. It is important to explain the purpose and techniques of work study to the people in top management.

Still more serious situation may arise if the attitude of the management is to use unethical practice. It may happen that the management may misuse the benefits forthcoming from the suggestions and scheme made by work study department. It has happened in past in many industries that with the application of new working methods the incentive plans introduced only to extract maximum from workers and subsequently retrench unwanted staff. This apparent gain which management thinks of getting by using such techniques, is only short lived and the huge advantages which could otherwise have obtained from the application of work study principles are lost for ever.

A good management on the other hand would encourage suggestions from the work study department, study the feasibility of implementing these suggestions from technical and economical aspects. Application of work study principles in one department or section can start a chain reaction throughout the organization ; in all other departments or sections. The investigation and improvement due to work study will spread in all directions throughout the company. The senior worker may be made to feel like novices when they find that new workers trained in the new methods soon surpass them in output and quality. In this case management would inform their staff about plans and programmes to be implemented and the worker and their representatives will be suitably informed through the work study personnel, supervisory staff and the foremen ; thus getting confidence, maximum co-operation and good support from all categories of staff.

3.5. WORK STUDY AND MIDDLE MANAGEMENT

Foremen or supervisors are the link between management and the workers. In the application of work study, the role of this middle management is very important. It is very difficult for the work study specialist to deal with workers without co-operation of supervisors. Their position in an organization is unique. The success of application of work study depends upon the attitude of supervisors. Before the work study specialist starts his work, the whole purpose of work study and the procedures involved must be very carefully explained to the supervisors. They should be taken into confidence. Unless this is done, the supervisors are likely to be difficult, if not obstructive, for many reasons.

Supervisors are normally selected on the basis of seniority from among the best-skilled persons in the enterprise. They are often middle-aged. They feel that the work for which they are responsible for years is being challenged. They feel that their prestige in the eyes of their boss and the workers will be lessened.

In many firms where work study specialist have not been used, the supervisors are responsible for the whole running of a plant. They carry out work planning, job developing, setting piece rates, hiring and firing workers. The some of these responsibilities have been taken away is likely to make him a loss of status.

Supervisors are the persons who understand the problems of workers. They are very close to workers. If any disputes arise between workers and work study person, supervisors are the first people who will be called upon to solve the problem.

As the supervisors are experienced persons and nearer to the practical side of the job than management, they are able to help in the selection of jobs to be studied and to understand the factors involved. Thus the work study specialist should retain the supervisors friendship and respect. He must never give a direct order to a worker. Any instructions must be given through the supervisor. For workers, non-technical questions should always be referred to their supervisors. Job to be studied should be selected through the supervisors. The work study person should discuss all technical matters, connected with the process, with the supervisors.

3.6. WORK STUDY AND THE WORKERS

The human being is one of the most crucial elements in any organization. Workers are most difficult people to deal with. The lot of time is wasted by workers in different ways. A lot of time is wasted due to absenteeism, late coming, starting the work late, and preparing for going early. Improper operation of machines results in high rate of rejection. Time is wasted due to accident also which occurs due to not observing safety precautions. Thus the time is wasted due to the behave of worker. Behavioural scientists believe that individuals are motivated to act in a certain way by a desire to satisfy certain needs. These needs arranged in a hierarchical pattern. According to Abraham Maslow, it is only when one need becomes satisfied that the next need in the hierarchy will start to exert its motivating influence.

The first need of the hierarchy is physiological need. This is basic need. Until this is satisfied, one will not be concerned with any other needs. The next need in the hierarchy is security. Security means protection against physical and psychological harm. After satisfying these two needs, the next motivating factors are affiliation, recognition and fulfilment. Affiliation means desire to belong to a group and to associate with others. The last needs expresses the desire of workers to get opportunity to show their talents. In practice, majority people are not satisfied with all these needs. They are partly satisfied with some of these needs and are left with some that are unsatisfied. In raising the productivity of an enterprise, there should not be placed too great an emphasis on labour productivity. The work study person should be open and frank in dealing with workers. He should not attempt to hide what is being done. Workers' representatives should be kept fully informed of what is being studied, and why. Asking for a worker's suggestions and ideas implicity serves to satisfy his need for recognition. The work study person has to study work, and not the workers. The objective of the application of work study is not merely to increase productivity but also to improve job satisfaction.

4

Working Environment and Safety

4.1. INTRODUCTION

Working environment plays an important role in raising the productivity. Bad working environment leads to occupational accidents. This has economic as well as physical consequences. The economic aspects includes direct costs and indirects costs such as medical care, compensation, material damage, cost involved in legal procedure etc. When accidents occur, the working time lost by the injured person, time lost due to other persons involved to help the injured person, production stops, etc. are the indirect cost involved. Physical consequences includes occupational diseases.

This leads to reduction in productivity and the increase in production rejects and manufacturing waste that result from fatigue due to bad working conditions. The productivity can be improved by improving the conditions under which people work. Consideration should be given to occupational safety and health in modern management techniques.

The hazardous working environment be a direct cause of occupational accidents and diseases. The workers dissatisfaction with working conditions may also be at the root cause of a decline in productivity.

4.2. OCCUPATIONAL ACCIDENTS

The occupational accidents occurs due to workers own fault. The dangerous accidents occurs due to quite ordinary actions of persons like stumbling, falling, the faulty handling of goods or use of hand tools, or being struck by a falling object. An analyst of accidents claims that 98% accidents could be easily avoided. Provided due precautions and care were taken well in time.

Improper acts result in violation of safety rules and non-observance of safety precautions. The safety rules such as do not grind without goggals, do not wear flopping clothes, 'wear shoes during forging or working in foundry shop,' should be followed by people themselves. Other causes of accidents are improper or defective tools or their unsafe or improper use, unsafe working position etc. also the violation of safety rules.

Technical progress has created new health hazards. An accident is the result of a technical, physiological and psychological factors. It depends on both the machines and the environment like lighting, noise, vibration etc. The first precaution may be taken to avoid accidents, is the elimination of potential causes, both technical and human. There are the observance of technical rules and standards, careful supervision and maintenance, safety training for all workers, and the establishment of good working relationships.

About 30% of all accidents occur in manual handling. Work study can contribute to reduce these accidents simply by reducing the number of handling operations and the distance involved in transportation of goods.

4.3. INDUSTRIAL ACCIDENTS

The accidents due to fire and explosion are very dangerous particularly in chemical industries. The prevention of fire, in this type of industry should receive particular attention. Generally there are safety teams leading by a safety managers. They are arranging safety week during which they arrange lectures on safety by expert peoples. In certain industries where a fire may lead to widespread material damage and, should it occur during working hours, to injury and even death.

The first principle of fire prevention is to design buildings, processes and storage facilities in such a way as to limit the possible combination of oxygen, fuel and an increase in temperature. Fire risk exist in all workplaces. The highest priority should be given to locating the fire as soon as possible.

The second principle is to eliminate or reduce sources of heat, thus limiting the rise in temperature. Measures may be taken such as restricting open flames *e.g.* welding toarches, and banning the smoking of cigarettes.

Management should be prepared for fire emergencies. Every work-place should have an emergency plan with information of the role of every worker in case of a fire. There should be at least two clear exists leading to safe area. All should know the location of these exists. There must be a way to communicate in the case a need arise to evacuate. This may be an alarm system or a visual signals such as flashing or revolving lights where there is a high level of noise. The right type of fire extinguisher should be provided in adequate numbers for the given risk, and placed accordingly. Every worker should be acquainted with the proper use of the extinguisher, including both when and when not to use it. There may be automatic fire protection provision such as sprinklers.

In developing a plant layout care should be taken to isolate the operations which is hazardous. Work place should be at high level from the ground. There should be a windows having a surface area not less than 17 per cent of the floor area. Minimum ceiling height should not be less than 3 metres. Enough air circulation should be their in the plant. To prevent accident, it should be ensure that each worker has an adequate minimum free-floor area which should not be less than 2 square metres per person. Walls and ceilings surfaces should have a finish which prevents the accumulation of dirt, avoids moisture absorption and reduces noise transmission.

4.4. GOOD HOUSE-KEEPING

Good housekeeping, when used with reference to a factory or workplace means to keep all the spare tools, raw materials, semifinished goods, and handling equipment at proper place where it does not obstruct the people. The tools and equipment should be returned to their designated place after use. Small hand tools should be displayed on board. Unnecessary items should be disposed of. Gang ways and passages should be kept clear. The work area should be kept clean. Surfaces should be free from oil and grease. Deposits of toxic material or chemicals may cause occupational diseases. Continuous cleaning of machinery and equipment can prolong their working life. Working clothes should be kept clean to reduce the skin diseases due to certain toxic substances or chemicals. Prolonged contact of the skin with certain chemicals may produce chronic dermatitis, which may be followed by the development of cancer. Workers working in chemical industries where they are exposed to dangerous or toxic substances should have washrooms with a tap for every three or four workers.

4.5. LIGHTING AND NOISE

Good lighting is required in doing a job in most of the industries. Good visibility of equipment, the product and data of the work process is an essential factor in accelerating production. In adequate visibility and glare may cause accidents. Visibility depends on a size and colour of the workpiece, its distance from the eyes, the lighting intensity, and contrasts of colour and lighting levels with the background. Lighting is more important in the case of precision work, or work in a dangerous environment. Lighting intensity should be varied according to the type of work. The level of illumination is measured in lux. The level of illumination also depends on the worker's age. Aged people require a higher level of illumination than young ones. Moreover, older people are highly susceptible to glare since their recovery time is longer. Accumulation of dust on light source, reduces the level of illumination. Regular cleaning of lighting fixtures is essential. Too high intensity of light may lead to development of visual fatigue and health disorders such as conjunctivitis and headaches. In mounting of general lighting units, care should be taken that the light uniformly diffused, glaring is avoided. Excessive contrasts between the worker's task and the general surroundings should also be avoided. The use of natural light should be encouraged. However day light varies with the season, the time of day, the distance of work stations from the windows, it is essential to have artificial lighting available all the time.

To get the enough lighting and to avoid glare, the following points should be considered for control of lighting :

- There should be an independent switch for the row of lighting fixtures closest to the window. This will ensure uniform distribution of light.
- For the precision work, localized lighting may be used to achieve the desired level.
- Highly shiny, glossy work surface should be avoided to prevent glare.
- Light fixtures should be cleaned regularly.
- Use diffusers to avoid direct eye contact with the light sources.
- The general lighting level should not exceed 500 lux.
- By suitable positioning of VDU, avoid glare.

Noise means any unwanted sound. The high noise level are annoying, but human beings get adjusted to them. The practical unit for measuring noise is decibel (dB). Noise may cause sensori-motor, neuro-vegetative and metabolic disorders. For the workers, machine noise may be a pleasant sound. But the persons resides near by the industries for them. Same sound may be noise. So the definition of noise varies with liking of people.

High levels of mechanisation, increased speeds of machines, the high density of machines at the workplace and nuisance factor of noise have resulted.

The human ear responds sounds of different frequencies. The unit of frequency is hertz (Hz) and the ear responds over the approximate range of 20 to 20000 Hz. The loudness of sound, as judged by the human ears depends on frequency as well as level. The ear is less sensitive to low and very high frequencies than to the middle range of frequencies from 1000 to 8000 Hz. In order to control noise, attempt have been made to locate the source and apply measures at the source to reduce noise. Foundations of motors and engines on vibration isolation pads, installation of silencers in the exhaust systems, use of acoustic wall materials are some of the methods used to reduce noise level. Noise impedes sound communication. It causes acoustical masking effect which reduces the intelligibility of speech. Background noise may hinder communication or may cause accidents. Normally sound level should not exceed 60—70 dB at a normal distance. Particularly in a intellectual work or work requiring intense concentration, the noisy environment may cause occupational deafness. Intermittent noise from rams, riveting hammers or large mechanical processes is disturbing and fatiguing. Prolonged exposure to noise above certain levels causes permanent damages to hearing and results in deafness. The exposure to continuous noise levels of 90 dB or above is dangerous to hearing, but 85 dB is a warning level which should not be exceeded.

To control noise, the first attempt to be made is to reduce the noise at source. This can be done by replacing noisy machines by less noisy ones.

The second attempt to be made is to prevent or reduce noise transmission by the installation of noise absorbent barriers between the noise source and the workers.

If the workers are exposed to a noise level of 90 dB for eight working hours, than the duration of noise exposure should be reduced acceptable limit.

There are some personal noise protection instruments such as ear-plugs or earmuffs are available which reduce the noise level up to 20 dB. Though the personal noise protection should be considered as provisional remedy until the workplace is permanently modified.

4.6. ENVIRONMENT

The human body maintains the necessary thermal balance by continuous exchange of heat with the environment. During physical activities, metabolic values varies. Normally it will be very high value than encountered at rest. Too high heat or too low temperature at the work place may affect on the body metabolism. The heat that the body is continuously producing must be dissipated when work is being done and absorbing heat from a high temperature environment. It is essential to avoid excessive heat or cold so that the body can maintain a thermal balance.

There are some working places where sources of heat such as furnaces, kilns or hot processes are present which produces excessive heat. At such places workers may have to deal with excess heat. In such working environments sweating is the only way to lose heat. As the sweat evaporates, the body cools.

Cold environment is more dangerous for survival owing to a drop in body temperature. The loss of body heat is uncomfortable. Long exposures in cold environment or extreme cold leads to death also. There are some places where workers have to work in the open in cold weather, such as construction work, electrical line or fisheries.

Workers in cold climates should be well protected against the cold by wearing suitable clothes, foot wear, gloves and a hat.

The another ways are, work. Periods in cold and hot environments should, where necessary, be altered with work periods in normal temperature. Workers should frequently take hot or cold drinks respectively to maintain a thermal balance.

Excessive humidity is also not tolerable in combination with cold environment. Relative humidity should be kept within a range of 40 to 70 per cent. Less humidity also create problems.

Excessively dry air can be a cause of respiratory tract diseases. There are many ways of controlling the thermal environment. Personal factors such as clothing, nutrition, personal habits, age also need to be taken into considerations in the attempts to attain the thermal comfort of workers.

4.7. ERGONOMICS

Scientific study of the relationship between man and his working environments is usually known as ergonomics. The international labour office defines ergonomics as "the application of human biological sciences in conjunction with the engineering sciences to achieve the optimum mutual adjustment of man and his work, the benefits being measured in terms of human efficiency and well being". The aims of ergonomics are to enhance functional effectiveness and maintaining or improving human welfare. Human factors should be taken into consideration in designing of machines and processes. The function of ergonomics is to develop the most comfortable conditions for the workers to reduce the physical work load ; to make the handling of machine levers and controls easier, to make better use of spontaneous and stereotyped responses. Sufficient research work has been carried out by physiologists and psychologists.

The interface between the worker and machine is very important. As the worker getting information from displays of the machine at their eyes or ears, his hands and legs acting accordingly. There is a variety of visual and auditory displays. The main aim of the displays is to establish a communication link between the machine and the worker. The quality and location of displays is very important. Electronic displays require specific guidelines for their selection, installation and use. There are varieties of displays as shown in Fig. 4.1. Quantitative displays are used when actual numbers are essential information may be obtained from a real display and from an artificial display. Scale and dial designs should be such that it can be read quickly and accurately. Location of the displays also concerned with the habit of the people. Well known example is the rear break of the bicycle. It is normally operated from left hand side handle. Peoples are normally habituated. So, rear break operating lever is always given at the left hand side. If there are number of dials to be read, the normal value in all the dials should be arranged at top. The normal value should be shown by red or green colour. So worker or operator can read

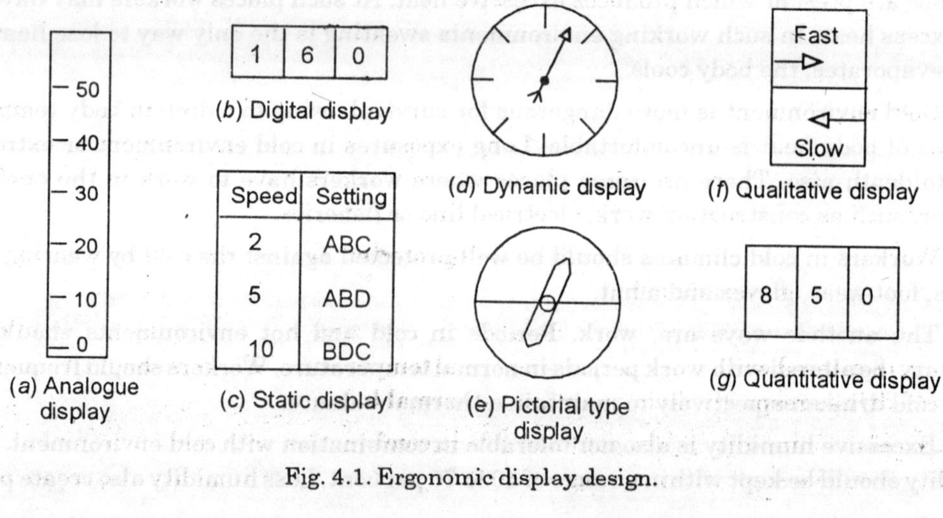

Fig. 4.1. Ergonomic display design.

the value which is below or above the normal value. The increasing values in the dial should be from left to right or from bottom to top. In the case of round dial, this increasing mode should be clockwise. The shapes, sizes and colours of control knobs are also important. It should be easily distinguishable from each other. Normally emergency switch should be easily visible and should be 'Red' in colour and larger in size so that it can be a real emergency switch. Mutually related dials and controls should be grouped together. Chapanis, Garner and Morgan give the following guidelines on dial design :

- A dial should be about 70 to 80 mm in diameter. This size of dial can be read at a distance of 750 mm or less.
- Marks should be located at the 0, 5, 10, 15, 20 etc. The marks at the 0, 10, 20 positions should be longer than those at the 5, 15, 25, ... positions.
- The separation between scale marker should be the same all around the dial.
- Values on the scale should increase in a clockwise direction.

4.8. FATIGUE IN INDUSTRY

It is difficult to define fatigue precisely. It is a result of physiological and psychological disturbance. Fatigue will lead towards inactivity. It affects the physiological functions such as

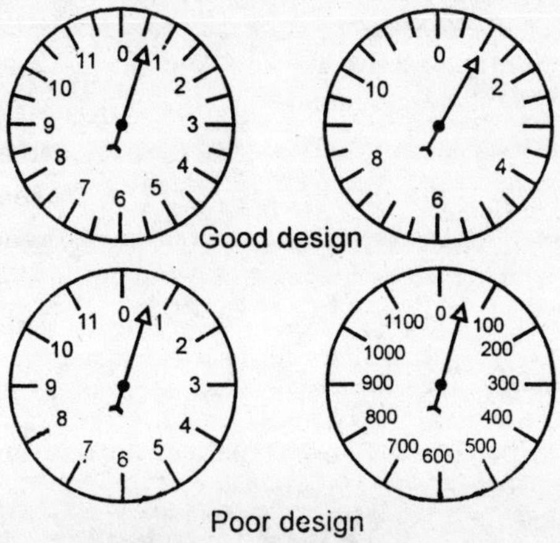

Fig. 4.2. Dial Designs.

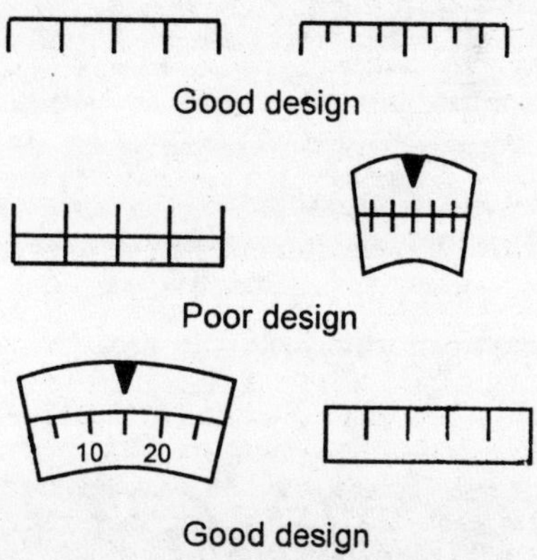

Fig. 4.3. Scale Designs.

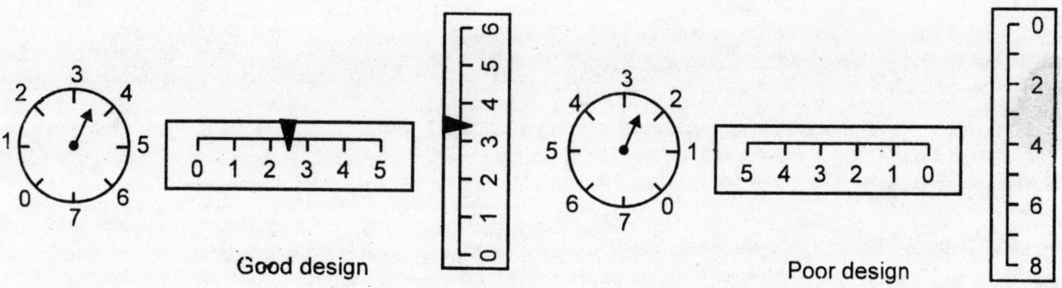

Fig. 4.4. Stereotype Displays.

heart rate, blood pressure, oxygen consumption, chemical composition of blood and urine, body temperature and rate of sweating. These may be used to measure fatigue. Adequate rest pauses should be given between the two operations so that speedy recovery is complete. The individual fatigue can be reduced by

(*i*) Improving the environmental conditions to the best possible level.

(*ii*) Designing the machines, tools and equipments for maximum efficiency with minimum physiological stress.

(*iii*) Selecting workers on the basis of physiological fitness for a given job.

The effect of temperature, light, noise and humidity on production, quality levels and physiological functioning of human beings is considerable and already discussed earlier.

5

Plant Layout and Materials Handling

5.1. INTRODUCTION

Plant layout is an important phase to be undertaken when planning a new industrial concern. As scientific management grew into industrial engineering, plant layout grew along with activities such as time and motion studies, cost accounting and production control. Layout work draws information from many of the other areas of industrial engineering.

About seventy years back, the machine tools were not developed as they are today. Machines tools were regarded as an addition to skilled hands of craftsman and hence the proper location of the machine tools and equipments were not given importance. Foreman was mostly responsible for properly locating various machines and equipments in the workshops. The persons working in industries were skilled artisans. The work carried out was mostly dependent upon their skill and intelligence.

With the implementation of a series of five years plans, the industries developed at a faster rate on the mass production of goods became a routine. High production machines were available in various industries and the plant layout became a fundamental function of efficient production planning.

It was observed during the study that the amount of time devoted towards method study and work measurement will be futile if considerable time is going to be ultimately wasted in handling and transportation of materials during the manufacturing of any product. By taking a cross section of the industry it was seen that between 15% and 85% of the manufacturing cost of an article was accounted towards materials handling alone. Some times as much as 50 tonnes of raw material was handled to produce one tonne of finished product. A good plant layout and work station layout is therefore, essential for economical production of goods.

Plant layout is the plan of an optimum arrangement of an industrial facility. Plant layout embraces the physical arrangement of various departments, machines, equipments, and services of men for economical, efficient and effective functioning while planning the production of any goods. It is a systematic arrangement of machines, equipments and work stations into and adjoining the plant buildings so that it will permit an uninterrupted flow of materials with minimum of handling in processing from raw material to finished product. It is possible to attain higher production by a well designed plant layout reducing wasted labour and incorporating efficient flow of material between work stations. Therefore, plant layout is one of many activities required for the effective overall design of an enterprise.

Problems of plant design are heavily dependent upon make or buy decisions. Obviously the plant design problem for a product which is to buy is not existent.

The determination of unit costs is usually the first step in a make-buy analysis. In such an analysis management is interested in : (*i*) reducing unit material and processing costs ;

(*ii*) minimizing costs investment, which demands consideration of both inventory control and optimum use of facilities ; and (*iii*) improving the product mix. The dicision to make or buy a product is essentially an engineering economy problem.

The size of a plant is dependent upon the volume of output proposed for it. Because the data are critical, an accurate forecast of the production requirements is essential in determining the size of the plant. Management must make this decision based on the best information it can obtain of the long run appraisal of the future scale of operation.

The break even chart is helpful in determining plant size. This chart indicate graphically, by plotting number of Rs. versus production capacity, the number of units that a concern must produce in order to break even.

5.2. THE PLANT LAYOUT

Plant layout has become one of the most interesting and important phases of industrial engineering. Industrial engineering is concerned with the design, improvement, and installation of integrated systems of men, materials and equipment. Plant layout is closely concerned with the design and installation of systems of men, materials, and equipment. Good plant layout is fundamental to the operation of an efficient industrial organization. Plant layout is a plan of, or the act of planning, an optimum arrangement of industrial facilities, including personnel, operating equipment, storage space, materials handling equipment, and all other supporting services, along with the design of the best structure to contain these facilities.

5.3. TYPES OF LAYOUTS

There are three basic methods of clustering an organization's activities. In the one method, all the equipment, and personnel engaged in the production of a single good or service are placed in a single department, the result is referred to as a product layout (line layout).

In the second method, all the equipment and personnel engaged in a unique process, one which is carried out in the course of producing many goods or services, are placed in a single department ; the result is referred to as a process layout (functional layout).

In the third method, all the equipment and personnel engaged in a process are brought to a product that is fixed in one position owing to its size ; the result is referred to as a fixed position layout. Extremely heavy and bulky products such as large ships, transformers, large vessels etc. possesses great problems regarding transportation and economic handling of components during manufacture. The fabrication and erection and erection of such custom built products is done on the fixed position. Manufacture of such product is a highly specialised job where accurate planning and careful judgement regarding the selection of men and manufacturing processes is to be exercised to achieve targets fixed for completion of product.

5.4. PRODUCT LAYOUT

This is also called line layout. When the product is produced in mass, this type of layout is used. This type of layout is adopted in an industry whose production is continuous.

In the product layout the equipments or machines used in the manufacture of a product are lined up in order of appearance. The machines are arranged according to the sequence of operations involved in manufacture. The raw material enters at one end of the line and goes from one machine to the next quite rapidly (within a department) with a minimum of work-in-process storage and materials handling. Fig. 5.1 shows the product type layout for the components 'A' and 'B'. As the material has to move minimum distance, the material handling cost

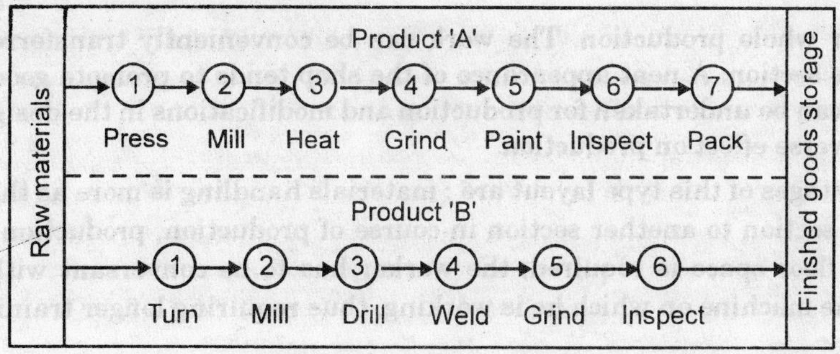

Fig. 5.1. Product Layout.

is reduced, manufacturing time is reduced, the production control is simple and the cost of articles produced is low at high volume production. The floor space required is minimum for this type of layout, operator training is economical and semi-skilled operator can be employed.

There are some disadvantages of this type of layout. It restrict its use for mass production only. Layout is less flexible so any changes in product design can not be easily incorporated. Due to duplication of several machines and equipments, the initial investment is very high. It requires provision of spare machines in case of failure of a machine. The breakdown of any single machine may shutdown a whole production line.

5.5. PROCESS LAYOUT

Process layout is also known as functional layout. In this layout, machines or processes of similar facilities are grouped together. In such a layout, the work usually moves through a series of departments and rarely from one work station to another within a single department. A more satisfactory flow of work can be obtained with a process layout, in which lathe will be grouped in one section, milling machines in second section and similarly other machines will be grouped in respective sections. This type of layout is shown in Fig. 5.2. This layout is adopted for low and medium volume production. This type of layout is generally preferred for general purpose repairs or small scale production units where jobs to be handled very frequently.

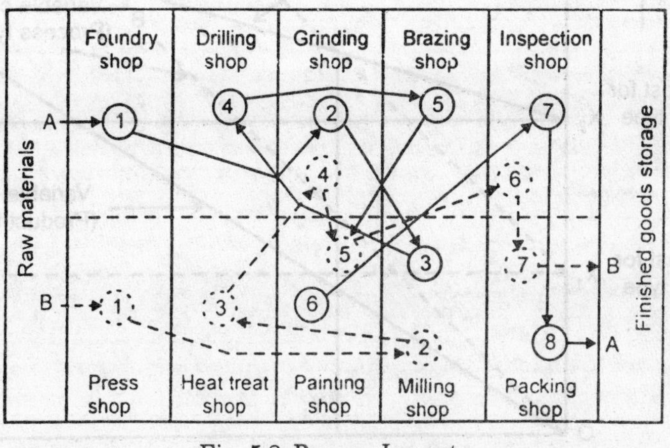

Fig. 5.2. Process Layout.

In this type of layout, operations are performed area wise. All stamping is done in the press section ; all milling operations are performed in milling section ; all grinding operations is done in grinding section and so on. If the product is not standardized, the process layout is more desirable. This layout is more flexible than product layout. Breakdown of a single machine

does not hinder whole production. The work can be conveniently transferred to any other machines in the section. A neat appearance of the shop tends to promote good housekeeping. Any type of job can be undertaken for production and modifications in the designs can be made without any adverse effect on production.

Disadvantages of this type layout are : materials handling is more as the product has to move from one section to another section in course of production, production control is more complex, more floor space is required, the worker has to be conversant with all operations pertaining to the machine on which he is working, thus requiring longer training period.

5.6. MIXED OR GROUPED LAYOUT

Mixed or grouped layout is a combination of product and process type layouts. In this type of layout advantages of both product and process layout is obtained. With the industrial development, this layout is widely used. This type of layout is used by the industries which have their own product and undertake restricted job work.

The selection of process type of layout or product type layout depends upon the initial fixed cost and the variable cost of producing parts. The break-even analysis can be used to select the type of layout. Fig. 5.3 is illustrated the break-even.

5.7. GROUP TECHNOLOGY LAYOUT

Group Technology (G.T.) is a recent innovation. In this similar types of products are grouped in to families. Here an attempt is made to reduce some of the complications of batch production. Products are grouped as per the manufacturing similarity. The product size and

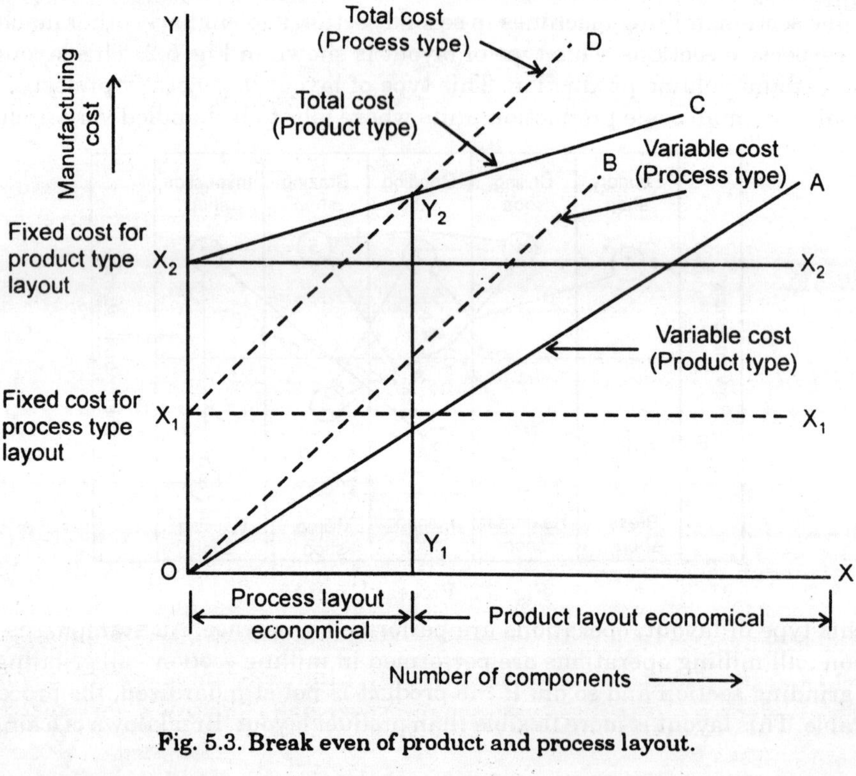

Fig. 5.3. Break even of product and process layout.

shape may be different but the operations to be performed may be similar on all the products. In the modern technique of group technology (GT), machine tools are grouped together according to the manufacturing requirements of a family of similar components. A whole factory might be sub-divided into groups, each group performing all the operations required on a family of products. These group layouts are known as GT cells. The advantages of GT layouts are reduced throughput times, improved control, greater flexibility, less work in process and higher worker morale.

5.8. MATERIALS HANDLING

No plant layout can be planned or analyzed without consideration of the materials handling problem. To convert the raw materials into finished product, it is necessary that movement of at least one of the three basic elements of production such as men, materials or machines take place. In most industries materials that moves rather than the men or machines. The materials handling system plays a significant role in the economics of manufacturing. It should be considered as a major factor in plant layout. The result of cost study have shown that each product may be handled as many as 50 to 60 times before it is ready for shipment. The cost of materials handling is as much as 20 to 25% of the total cost of converting the raw material into the finished product. Study of materials handling system includes movement of raw materials from stores, through production processes to the packaging of final product in the most efficient and safe manner so as to achieve higher productivity at a minimum handling cost.

There are many definitions of materials handling. The American materials handling society has defined it as the art and science involving the movement, packaging and storing of substances in any form. The substances means gaseous, liquid or solid form of materials. Throughout history, materials handling has been considered as an art and only fairly recent years scientific approach have been developed to materials handling problems.

The term materials handling itself indicates any movement of materials, whether the movement is vertical or horizontal or both. It also includes that the movement is by mechanical means or by hand.

There are varieties of materials handling equipments in use at present time for different purposes. The materials handling equipments classified according to functional activities. The materials handling hand book classifies these functional areas as follows :

(1) Bulk handling (2) Unit handling (3) Industrial packaging (4) Ware housing (5) Carrier handling (6) Handling operation analysis.

The materials handling equipments may be conveniently classified as under :

Major Classes	**Sub Classes**
Fixed path	Conveyors, hoists, lifts.
Limited area	Cranes
Large areas	Tractors, trucks, railway, aircraft, water carrier.

The sub classes have been developed for more specific types of application within the major classes. Various types of conveyors (belt conveyors, chain conveyors, elevating conveyors, and roller conveyors) are used where bulk of materials is to be moved continuously from one place to another. Cranes and hoists are used for intermittent movements of materials within a fixed area. Industrial trucks such as powerized hand lift truck, fork lift truck, motor truck

mounted cranes etc. are used where materials are to be moved intermittently through different routes and large distances (Fig. 5.4). For the planning of an efficient materials handling system the following points should be considered :

- **Machines and equipments should be arranged in such a way that the unnecessary movement can be avoided as far as possible.**
- **There should be least manual handling, mechanical means should be used wherever possible.**
- **The handling cost varies inversely with quantity of materials to be moved per tour. The cost decreases as the quantity increases.**

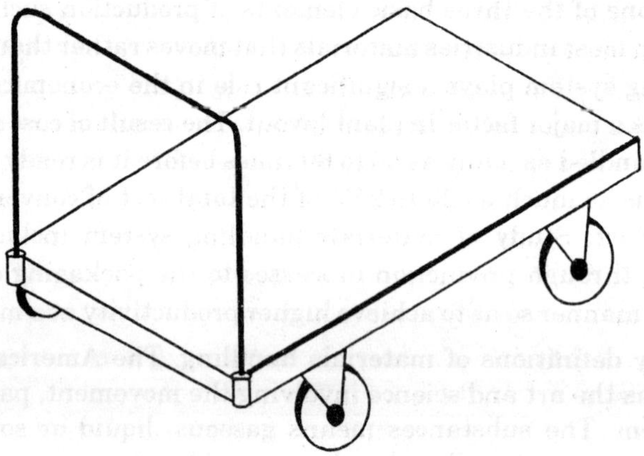

Fig. 5.4. (*a*) Platform trolley with single swivelling castor wheels.

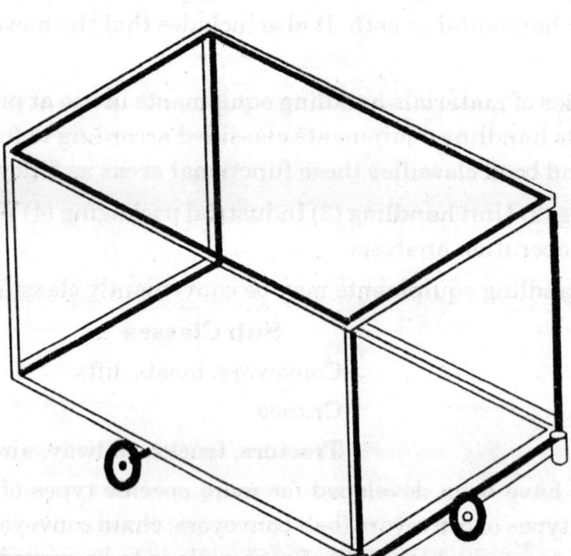

Fig. 5.4. (*b*) Cage trolley for carrying finished products.

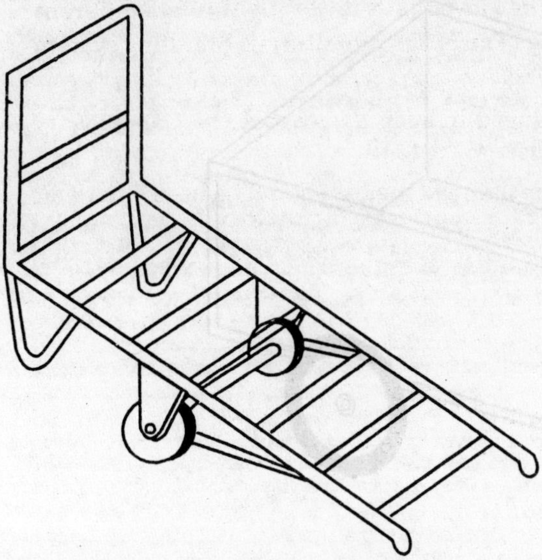

Fig. 5.4. (*c*) Tubular Luggage Trolley.

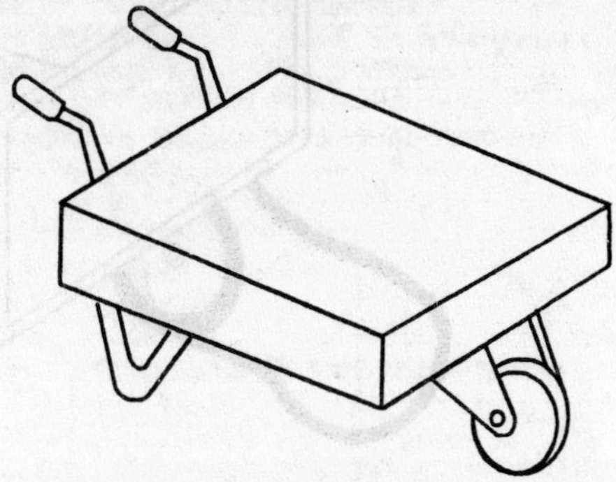

Fig. 5.4. (*d*) Pan Type Single Wheel Barrows.

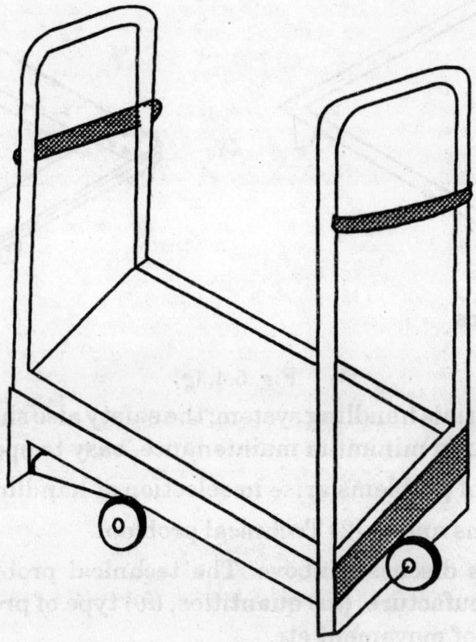

Fig. 5.4. (*e*) Double Handed Trolley.

- **Economy is obtained by using gravity chutes, empowered roller conveyors and versatile equipments. The versatile equipments can be used for various purposes, thus, idle time of handling equipments reduces. The materials should move by the shortest route in quantities to reduce cost.**

Fig. 5.4. (*f*) Tilting Hand Trolley.

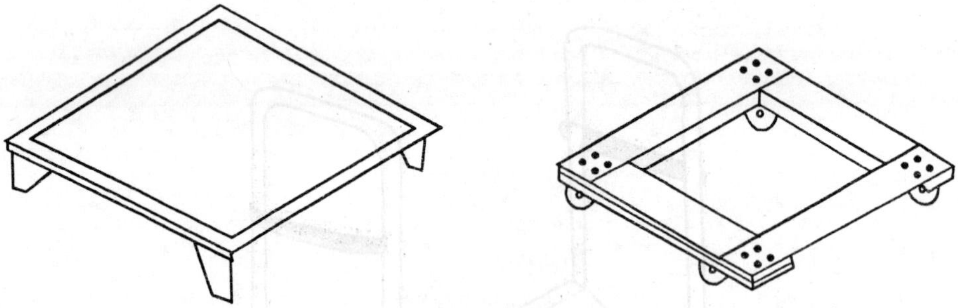

Skids Dollies

Fig. 5.4. (*g*)

- **In selecting materials handling system, the safety also should be considered. It should be safe and required minimum maintenance, easy to operate and economical.**

There are two types of problems arise in selection of handling equipments :

(1) Economic problems and (2) Technical problem.

Economic problem is discussed above. The technical problem includes (*i*) floor space required, (*ii*) method of manufacture, (*iii*) quantities, (*iv*) type of product to be handled, (*v*) path of movement, (*vi*) direction of movement etc.

Space requirements affect the plant layout. The aisle space requirements influence the selection of tractor and lift trucks.

Different types of materials are required different types of equipment to handle. Some handling equipment such as belt conveyors, can be used for both, bulk or packaged materials. Elevators can be used for bulk type material only. While roller conveyors are used for packaged type material only.

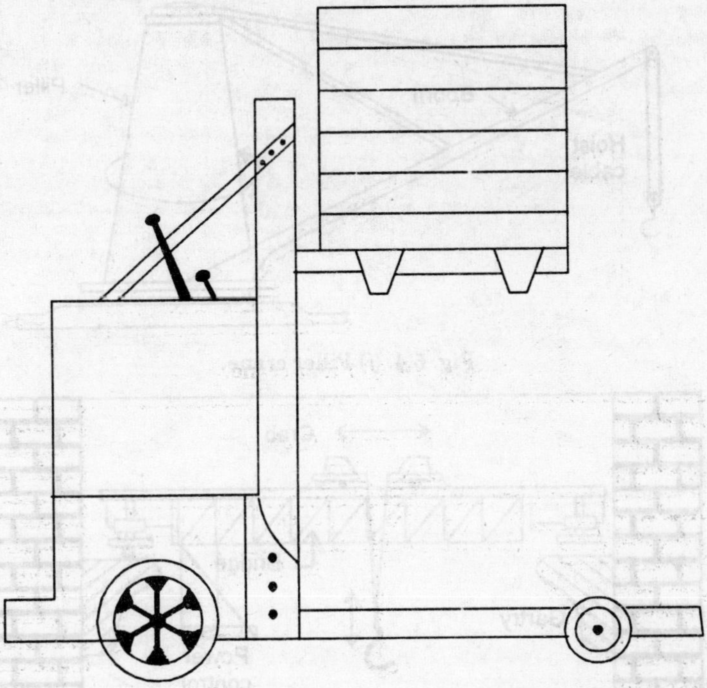

Fig. 5.4. (*h*) High lift elevating platform truck.

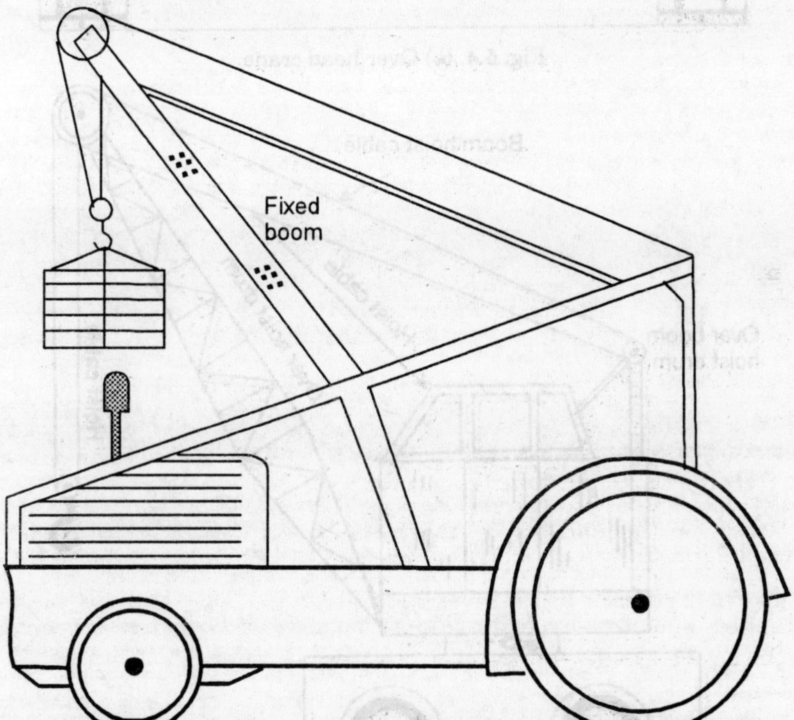

Fig. 5.4. (*i*) Rigid boom industrial crane.

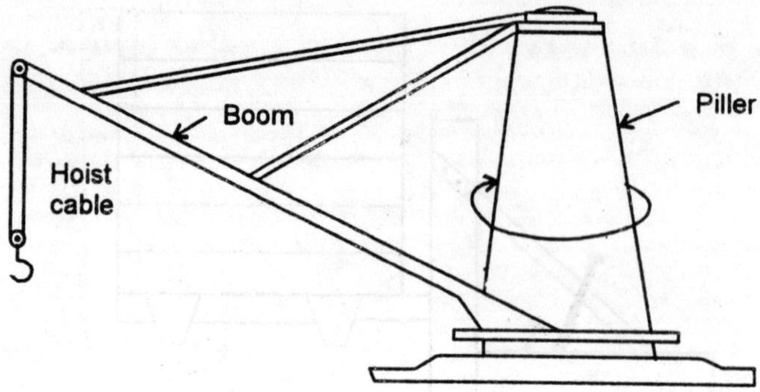

Fig. 5.4. (j) Piller crane.

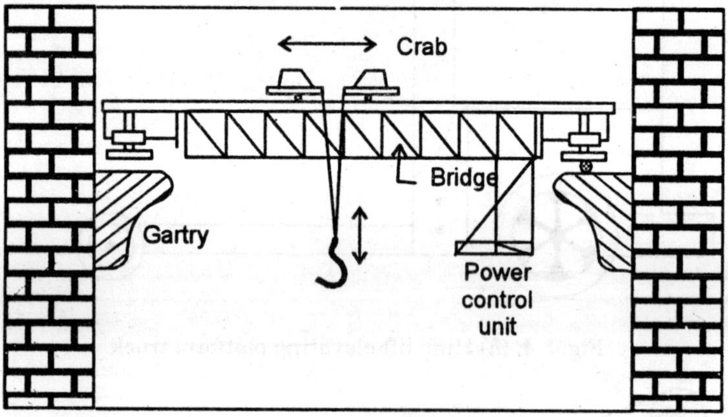

Fig. 5.4. (k) Over head crane.

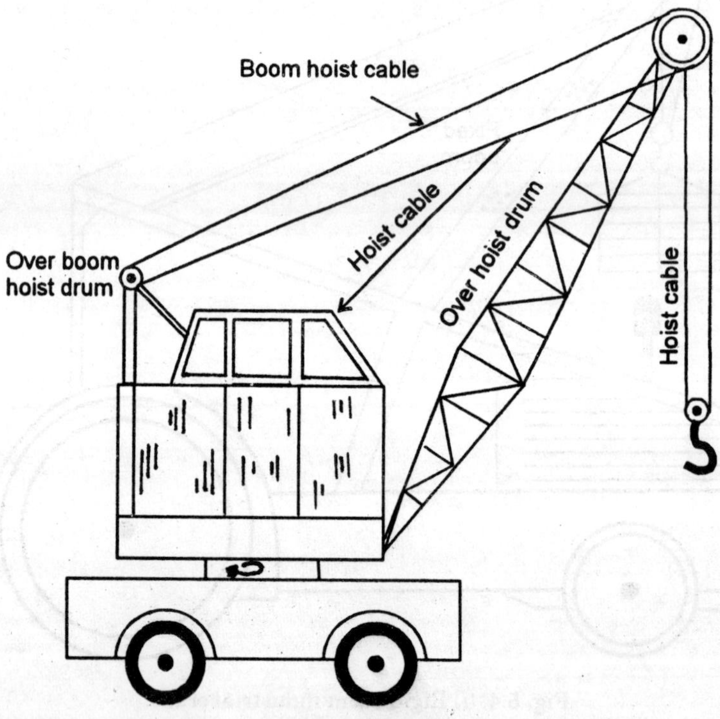

Fig. 5.4. (l) Crane truck.

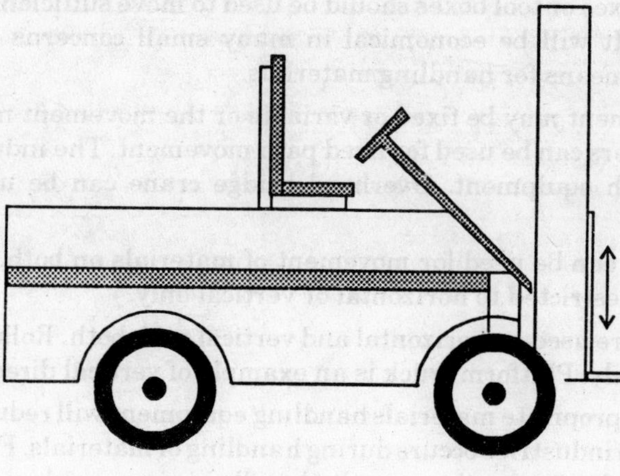

Fig. 5.4. (*m*) Power operated fork lift.

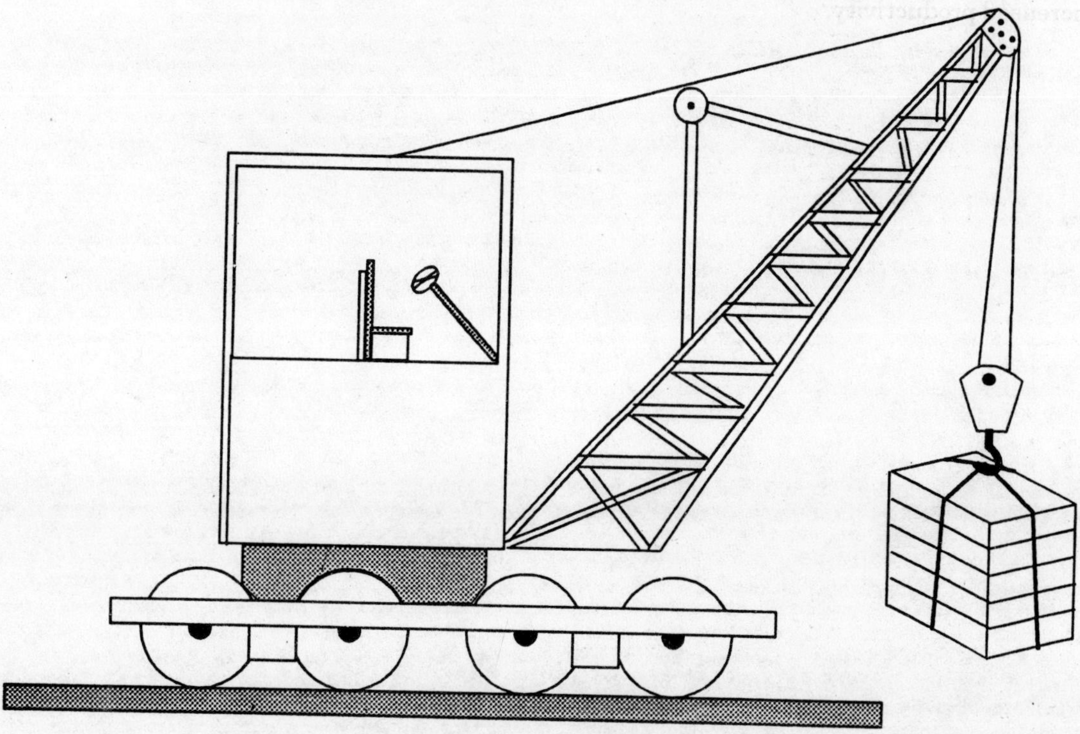

Fig. 5.4. (*n*) Locomotive crane.

In small jobbing shop the materials to be moved at any one time may be small but it may be heavy and may involve much labour even if only for a limited period. In such type of industry, it is found that a skilled machine operator lifting a large machine vice from the floor. In turning section, a skilled turner might be seen mounting a heavy face plate on lathe without any mechanical aid. The skilled labour is expensive to be used on a labouring job.

The metal work boxes or tool boxes should be used to move sufficient quantities from one place to another place. It will be economical in many small concerns to introduce several inexpensive mechanical means for handling materials.

The path of movement may be fixed or variable or the movement may be restricted to a limited area. The conveyors can be used for fixed path movement. The industrial lift truck is an example of variable path equipment. Overhead bridge crane can be used for limited area movement.

Some equipments can be used for movement of materials on both, horizontal path and vertical path. Some are restricted to horizontal or vertical only.

Screw conveyors are used on horizontal and vertical path both. Roller conveyors are used in horizontal direction only. Platform truck is an example of vertical direction only.

The selection of appropriate materials handling equipment will reduce the accidents also. About 43% of accidents in industries occurs during handling of materials. Further the movement of materials to be handled is faster, the quantity handled is increased and the cost of handling is reduced. The manual labour which causes fatigue is reduced considerably resulting in increased productivity.

6

Work Simplification

6.1. INTRODUCTION

There are two main techniques of work study namely work simplification (method study) and work measurement (time study). Work simplification is variously known as motion study, method study and methods engineering. It is defined as 'the systematic recording and critical examination of ways of doing things in order to make improvements'. Gilbreth was the originator of this technique. He had concentrated on the study of movement of human body and designated as motion study. After that the scope and application of this technique was widened tremendously. That is why this technique is now referred as work simplification.

The basic approach to work simplification technique consists of the following eight steps :

- **Select.** Select the job to be studied and define its boundaries.
- **Record.** Record all relevant facts about the job by direct observation.
- **Examine.** Examine the job critically and challenge its purpose, place, sequence and method of performance.
- **Develop.** Develop the most practical, efficient, and economic method of manufacture.
- **Evaluate.** Evaluate different alternatives to develop a new improved method.
- **Define.** Define the new method, as a result, in a clear manner and present it to the management, supervisors and workers.
- **Install.** Install the new process as a standard practice.
- **Maintain.** Maintain the standard practice by introducing control procedures.

6.2. SELECTION OF WORK TO BE STUDIED

Almost any job can be studied for an investigation and can be improved by proper application of principles of work simplification. There is always a scope for the improvement of any job. Work study principles can be applied for the study of an assembly at a single work station to making a layout of a plant for mass production. The study is not only confined to manufacture of engineering product only, but it can also be extended to other activities such as routine work in hospitals, serving of meals in a hotel, routine work in banks, packing of materials in a provision stores etc. Selection of job for work simplification is based on the consideration of the following factors :

1. Economic or cost effective considerations.
2. Technical considerations.
3. Human considerations.

(1) Economic Considerations. The economic importance of any job should be considered first. If a job is very small or it is not expected to run for long time, it is of a temporary nature, and having few persons working on it ; the time should not be wasted to start or to continue a long investigation. It should be questioned : "Will it pay to begin a method study of this job ?" or "Will it pay to continue this study ?"

The following points should be considered in selection of a appropriate job :

(A) The job should be selected whose production cost is very high or which is very high profit generating.

(B) The job should be selected which creates bottleneck situation in the industry. In bottleneck situation, there is conjuction of job at one work station and there is no work at the next work station. This is because of lengthy operations that consume a lot of time at one work station and holding up other production operations which are to be performed next.

(C) The job which requires repetitive work and involve great deal of labour should be selected.

(D) The job which is required to move long distances between two work stations or which require repeated handling of materials should be selected.

The job whose production cost is very high or which is very high profit generating can be identify by pareto analysis which is also called ABC analysis.

(2) Technical Considerations. This involve higher initial investment in advanced technology. These considerations depends upon the management policy. The introduction of new technology depends upon the type of information needed, effect on production cost, effect on profit etc. Some time it may happen that the production cost increases due to introduction of new technology. It is important to select a most efficient and practical process of manufacture commensurate with the function of the part. This should take into consideration the use of high productive machines, standardization of components and effective quality control of the product.

(3) Human Considerations. It is not possible to satisfy every one. There are some workers may be dissatisfied by certain operations which they have to perform. They feel fatigue or monotony or may be unsafe or clumsy to operate. If such operations are addressed by work study personnels as part of an over all work study programme, the benefits of work study can become more apparent to the work force. The work study personnel should see that proper working environment is provided. Steps are taken to uplift the workers' morale by way of regular payments. The safety precautions should be strictly adhered to, thus reducing number of accidents and loss of man-machine hours.

6.3. RECORDING THE FACT

Once the job has been selected for method study, the next step of the basic procedure is to record all the relevant facts about the existing method, the work station layout, and the equipments used for the production of that part. This is very important step, therefore, all the details are recorded very accurately. The success of improving the existing method or developing a new method depends on the accuracy with which the details are recorded, because these will provide the basis for critical examination and the development of the improved method. The record should be clear and concise because it serves as a basis for subsequent analysis and critical examination of the existing method.

The usual method of recording facts is to write down all the details step by step. This is very lengthy and tedious procedure, may run in several pages of closely written script even for simple job, which may take only a few minutes to perform the job than to record the details. No body will have a time or patience to read long description carefully and visualise the difference between existing and proposed methods of manufacture. This difficulty can be overcome by using other techniques or tools of recording such as charts and diagrams. In this charts or diagrams, detailed information may be recorded precisely in standard form and it may be understood by all method study persons, in whatever company or country they may be working. There are several types of charts available which are in standard format. There are two groups of these charts :

(1) used to record a process sequence which do not describe the events to scale ; and
(2) used to record events, also in sequence, but describe it on a time scale.

The most commonly used recording techniques are again classified as shown in Fig. 6.1. Diagrams indicates movements more clearly than charts can do. Diagrams usually do not show all the information recorded on charts, so they are used as a supplement rather than replace.

Recording Techniques

Charts using a time scale	Charts without time scale indicating process sequence	Diagrams indicating movements
* Multiple activity charts	* O.P.C.	* Flow diagrams
* Simo charts	* Flow process chart	* String diagram
	* Two handed process chart	* Cycle graph
		* Chrono cycle graph
		* Travel chart

Fig. 6.1. Commonly used charts.

6.4. PROCESS CHARTS

A process chart is a graphic presentation of the activities that occur during a production. These charts usually begins with the issue of raw materials from store and ends with the parts or assembly is completed and packed or deposited in store for subsequent reissue.

The recording of the facts about a job or operation on a process chart is made by using a set of five standard symbols. These symbols are recommended by the American Society of Mechanical Engineers in 1947. The detail description of the symbol is given below :

○ **OPERATION**. This is the main steps in a procedure. An operation occurs when an object or part is changed in any of its physical or chemical characteristics, is assembled or disassembled from other object. An operation is also occurs when information is given or received, or when planning or calculating takes place. Every operation takes the material, component or service one step near to its completion, whether by changing its shape or its chemical composition or by adding or subtracting material.

An operation is denoted graphically on process chart by a circle.

□ **INSPECTION**. An inspection occurs when an object is examined for quality or is verified for quantity during or after the process. An inspection does not take the object any near to completion. It only verifies that an operation has been carried out according to quality or/and quantity.

An inspection is denoted graphically on process chart by a square.

⇨ **TRANSPORT.** A transportation occurs when an object, worker, material or equipment is moved from one place to another, except when such movements are part of an operation or are caused by the operative at the work station during an operation or an inspection.

A transportation is denoted graphically on process chart by an arrow.

D **DELAY.** It is also called temporary storage. A delay occurs when conditions do not permit or require immediate performance of the next planned step. It is a delay in the sequence of events. It occurs when any object laid aside temporarily without record until required. For example a component turned on a lathe is stored in bins to be subsequently delivered to plating section.

It is denoted graphically on process chart by a capital letter 'D'.

∇ **PERMANENT STORAGE.** A storage occurs when an object is kept and protected against un-authorized removal. The issue of materials or tools from permanent storage is only possible after proper authorization. It indicates a controlled storage. A requisition, or other form of formal authorization is required to get an article out of permanent storage.

It is denoted graphically on process chart by a symbol ∇.

▢ **COMBINED ACTIVITIES.** The circle within the square indicates combined activities operation and inspection. These activities are performed at the same time or by the same operative at the same work station. Similarly other activities also can be combined such as operation and transportation. The symbol for this combination is arrow within the circle.

6.5. OPERATION PROCESS CHART

It is also known as the outline process chart. It is a graphical representation of a process. It is a "bird's-eye" view of a whole process or activity before starting a detail study. It gives an overall picture by recording in sequence only the main operations and inspections. It shows in chronological order, the principal operations carried out and the inspections made, time allowed and materials used in manufacturing processes, from raw material stage to the finished product stage. In preparing this chart, only the symbols for "operation" and "inspection" are necessary. In addition to the symbols and their sequence, a brief note of the nature of each operation or inspection is made and the time allowed for it is also added. An example of an outline process chart is given in Fig. 6.4 for the assembly of a vice. Figs. 6.2 and 6.3 shows the assembly and details of the vice respectively. It consists of a (1) Fixed jaw (main body) (2) Movable jaw (3) Bottom plate (4) Screws for assembling bottom plate with movable jaw (5) Lead screw (6) Grub screw to lock lead screw with movable jaw (7) Handle and (8) Collar fitted at the other end of handle.

In making an operation process chart it is usually convenient to start with a vertical line down the right-hand side of the page to show the operations and inspections undergone by the principal unit or component of the assembly. The time allowed per piece in hours is shown to the left of each operation. No specific time is allowed for inspections. The brief descriptions of the operations and inspections which would normally be shown alongside the symbols have been omitted so as not to cluster figure.

The brief descriptions of the operations and inspections for each component is given below :

- **Fixed jaw (Main body).** Made from C10 case hardening steel.

① Face on shaper (0.025 hours)

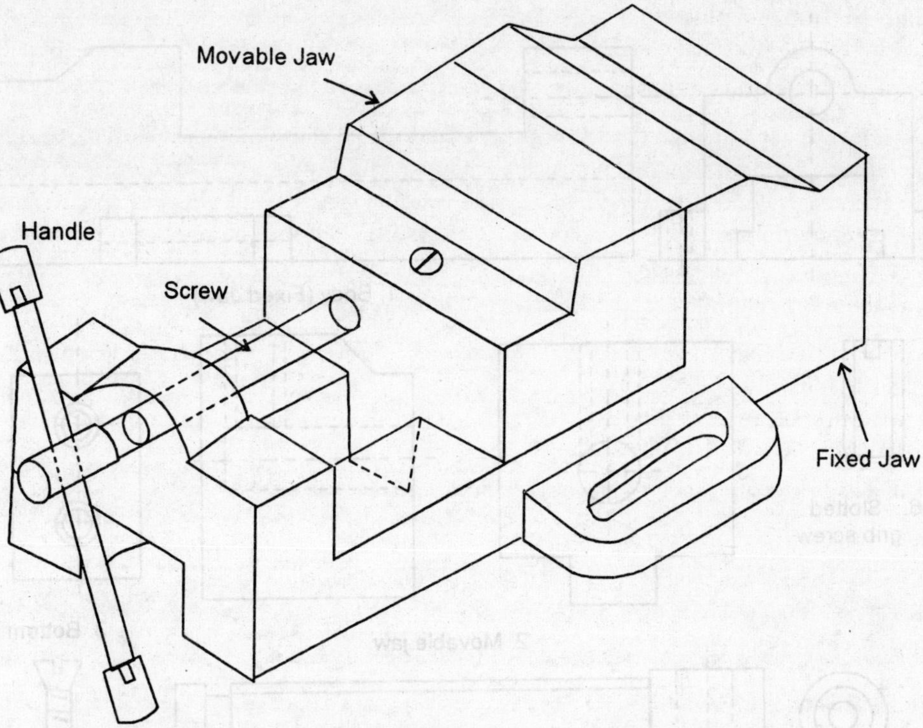

Fig. 6.2. Small Parallel Jaw Vice.

② Face opposite end on shaper (0.025 hrs)

[1] Inspect for dimensions.

③ Shaping on top surface (0.05 hrs)

② Inspect for dimensions.

④ Mill bottom surface (0.025 hrs)

⑤ Mill bottom slot (stradle milling) and slots for plate (0.05 hrs)

[3] Inspect for dimensions.

⑥ Boring on lathe (0.01 hrs)

⑦ Square thread cutting on lathe (0.03 hrs)

[4] Inspect thread for proper functioning.

⑧ Milling two holes by using end mill cutter on vertical milling m/c (0.05 hrs)

[5] Inspect for the dimensions.

● **Movable jaw.** Made from C10 case hardened steel

⑨ Shaping on top (0.03 hrs)

⑩ Shaping on both sides (0.05 hrs)

⑪ Shaping on bottom side (0.03 hrs)

[6] Inspect for dimensions.

⑫ Drilling blind hole (0.010 hrs)

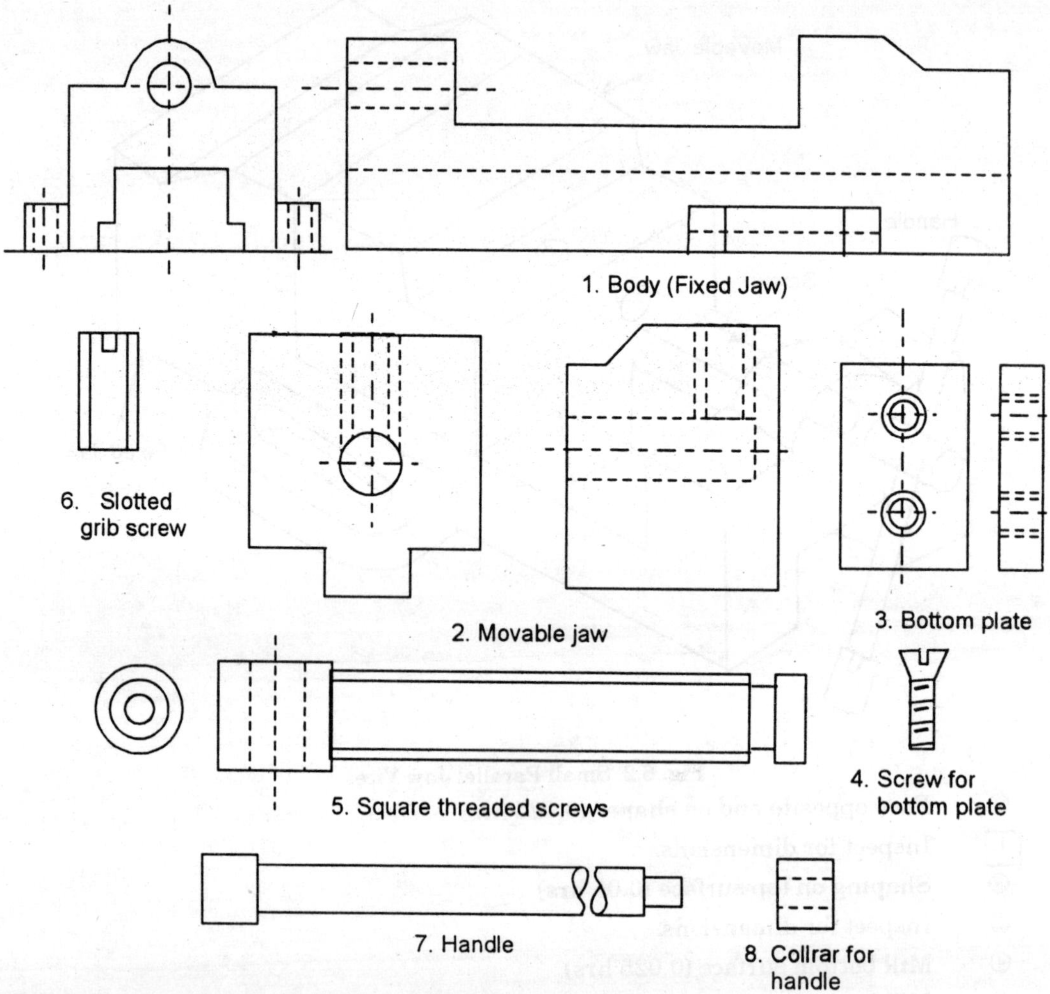

Fig. 6.3. Details of Vice.

(13) Drilling hole on top (0.010 hrs)

(14) Tapping (0.010 hrs)

[7] Inspect for dimensions.

- **M.S. Bottom plate.** Made of mild steel.

(15) Cut to size on hacksaw machine (0.02 hrs)

(16) Drilling hole : 1 (0.01 hr)

(17) Drilling hole : 2 (0.01 hr)

(18) Counter sinking (0.008 hr)

(19) Counter sinking (0.008 hr)

[8] Inspect for dimensions.

- **Screw for bottom plate**

[9] Purchase and inspect screws for bottom plate.

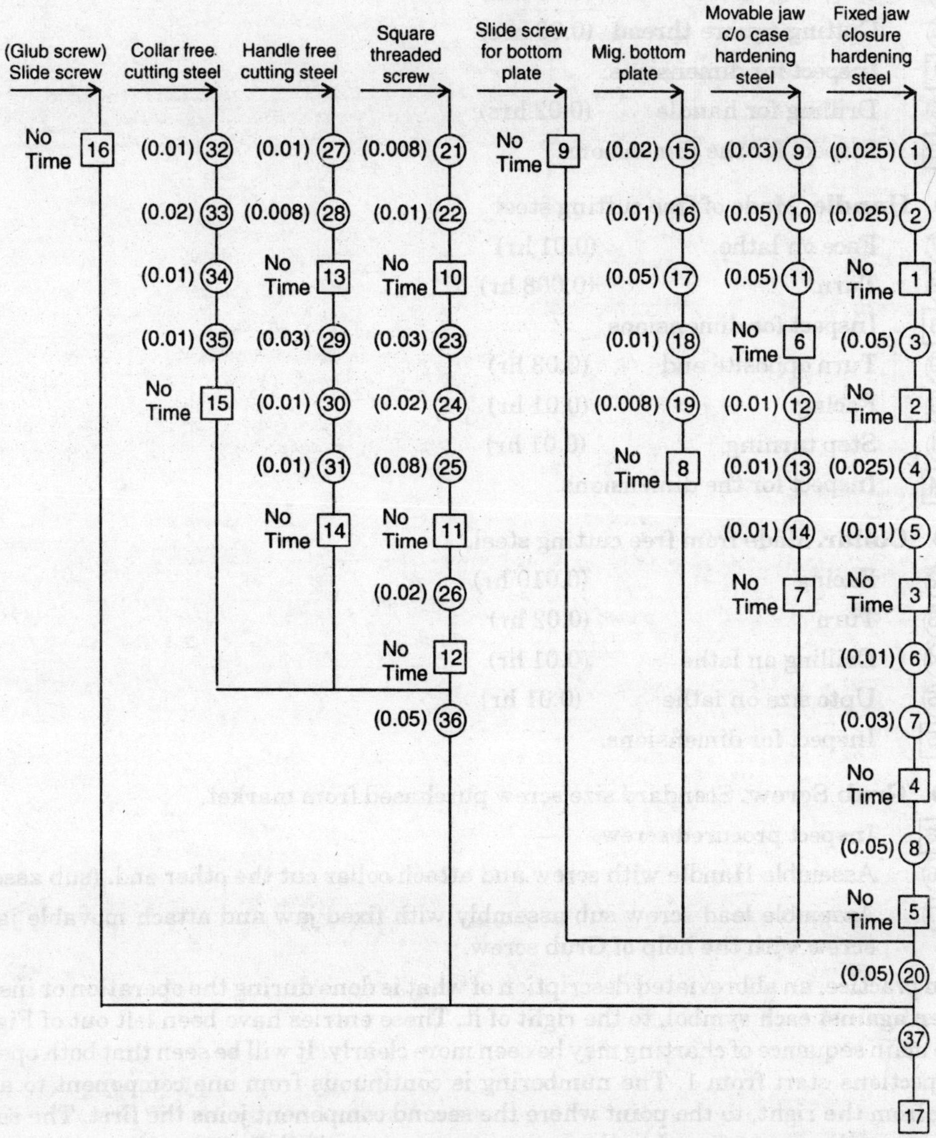

Fig. 6.4. Out Line Process Chart : Vice Assembly.

(20) Assemble movable jaw with fixed jaw with the help of bottom plate and 2 screws (0.08 hrs).

- **Square thread screw (Lead screw).** Made of C30 steel.

(21) Face on lathe (0.008 hrs)

(22) Turn on lathe (0.010 hrs)

[10] Inspect for dimensions.

(23) Turn other end (0.03 hrs)
(24) Grooving on lathe (0.02 hrs)
(25) Cutting square thread (0.08 hrs)
[11] Inspect for dimensions.
(26) Drilling for handle (0.02 hrs)
[12] Inspect for the dimensions.

- **Handle.** Made of free cutting steel

(27) Face on lathe (0.01 hr)
(28) Turn (0.008 hr)
[13] Inspect for dimensions
(29) Turn opposite end (0.03 hr)
(30) Facing (0.01 hr)
(31) Step turning (0.01 hr)
[14] Inspect for the dimensions.

- **Collar.** Made from free cutting steel.

(32) Facing (0.010 hr)
(33) Turn (0.02 hr)
(34) Drilling on lathe (0.01 hr)
(35) Upto size on lathe (0.01 hr)
[15] Inspect for dimensions.

- **Grub Screw.** Standard size screw purchased from market.

[16] Inspect procured screw.
(36) Assemble Handle with screw and attach collar cut the other end. (sub assembly)
(37) Assemble lead screw sub assembly with fixed jaw and attach movable jaw with screw with the help of Grub screw.

In practice, an abbreviated description of what is done during the operation or inspection is written against each symbol, to the right of it. These entries have been left out of Fig. 6.4 so that the main sequence of charting may be seen more clearly. It will be seen that both operations and inspections start from 1. The numbering is continuous from one component to another, starting from the right, to the point where the second component joins the first. The sequence of numbers is then transferred to the next component on the left and continuous through its assembly to the first component until the next assembly point, when it is transferred to the component about to be assembled. The assembly of any component to the main component or assembly is shown by a horizontal line from the vertical operation line of the minor component to the proper place in the sequence of operations on the main line. Fig. 6.5 shows some of the conventions used when drawing outline process charts.

Subsidiary component is shown by drawing a horizontal line, after inspection 3, with main part, and is assembled to it during operation 7. Assembly undergoes two more operations, number 8 and 9, each of which is performed four times in all, as shown by the "repeat" entry. Next number after these repetition is 18, not 10.

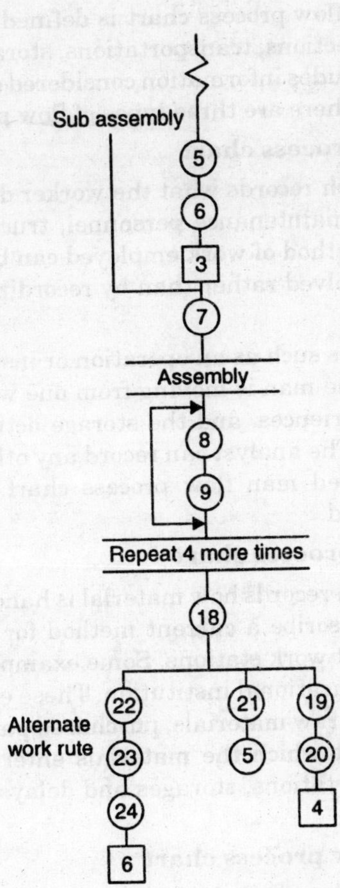

Fig. 6.5. Charting Convention.

As explained earlier, the operation process chart (O.P.C.) is the "bird's eye" view of the activities involved, for the purpose of eliminating unnecessary ones or combining those that could be done together. It is usually necessary to go into details greater than the OPC provides.

6.6. FLOW PROCESS CHARTS

A process chart is simply a unique way of describing a method of work. It is unique in the sense that with a description of this sort, an analyst is more likely to get ideas for possible improvements than he would with a description of some other sort.

A number of different kinds of process charts are available for use. The specific chart that will be employed in a method analysis will depend upon the nature of the task being studied and on the amount of detail the analyst wants to incorporate in his description of the task.

An operation process chart gives an overall picture of a process. But for economical production and proper layout more information regarding movements of men and materials is needed. This can be obtained from a flow process chart. A flow process chart contains considerable more information than the operation process chart.

According to A.S.M.E. a flow process chart is defined as "a graphic presentation of the sequence of all operations, inspections, transportations, storages, and delays, occurring during a process or procedure ; also includes information considered desirable for analysis such as time required and distance moved. There are three types of flow process charts :

(1) Worker type flow process chart

A flow process chart which records what the worker does. Some examples are the work of messengers, police officers, maintenance personnel, truck drivers, supervisors, and sales people. In all such cases, the method of work employed can be best described by recording the activities of the individuals involved rather than by recording what happens to the materials involved.

In this chart, the activities such as an operation or inspection is carried out by the man, a transportation occurs when the man is moving from one work station to another, a delay is something which the man experiences, and the storage activity is not considered because it cannot be performed by a man. The analyst can record any other information such as distances and required times. A completed man flow process chart is analyzed to obtain ideas for improvement in the work method.

(2) Material type flow process chart

A flow process chart which records how material is handled or treated. The material type flow process chart is used to describe a current method for a task in which "materials" are processed at a number of different work stations. Some examples are the production of a lamp ; registration of students at an educational institution. These examples indicates that materials involved might be such things as raw materials, purchased parts, reports, forms, or people. The chart will indicate the points at which the materials enter the process and then show the operations, inspections, transportations, storages and delays to which the materials are subjected.

(3) Equipment type flow process chart

A flow process chart which records how the equipments are used. Both of the process charts described above lend themselves for use when the activities to be described take place at more than one work station. Equipment type flow process chart is described a task which is carried out at a single work station. Equipments used at a single work station is described in this chart. In general, a work station is represented by the area occupied by the operator and the facilities with which that operator is provided. As an example, an office occupied by an administrative assistant. It is expected that the office will consist of a number of location at which various things are done. To illustrate, the assistant may go to where a storage cabinet is located in the office to obtain needed supplies, to where a typewriter is located to type a memorandom, to where a copier is located to duplicate a report and so on. These activities are described in the same terms as where used in the material type and worker-type flow process charts, that is, in terms of operations, inspections, transportations, storages, and delays. However, there is an important difference. In an equipment type flow process chart, a transportation is said to occur whenever material is moved or a man moves from one location to another location of equipments in the given work station.

A flow process chart is prepared in a similar manner to that in which the operation process chart is made, but in addition to the "operation" and "inspection" symbols, three more symbols "transport", "delay" and "storage" are used in this chart. It is customary to use the active voice of verbs for descriptions on worker type charts, and the passive voice on material type and equipment type charts. The following examples shows the typical entries on worker type and material type flow process chart.

Worker type	**Material type**
Shears plate	Plate sheared
Carries to drilling M/C	Carried to drill M/C
Picks up nut	(nut) picked up
Inspect for finish	Finish inspected

A flow process chart gives more detail than an operation process chart. It does not usually cover as many operations per sheet as may appear on a single operation process chart. It is usual to make a separate chart for each major component or an assembly, so that the amount of handling, delays and storages of each may be independently studied. This means that flow process chart is usually a single line. It records all factors pertaining to a process. It visualise the complete sequence of events in a process. This enables one to improve layout of the machinery and equipments ; thus reducing the movements of raw materials, finished products and men. A precise study and analysis of these records may further lead to reduce some operations and inspections, certain delays and storage and may suggest an improvement in the materials handling facilities.

The details which appear on a chart must be obtained from direct observation. Once they have been recorded on the chart, they remain available for reference and for explaining the situation to others. Charts must not be based on memory but must be prepared as the work is observed. Recorded details should be reviewed and confirmed with the supervisor. This confirmation ensures that the facts are correct and it strengthens the bond between the work study person and the supervisor, who appreciates that his opinion is valuable to the investigation. A high standard of neatness and accuracy should be maintained in preparing a chart. An untidy chart will make a bad impression and may lead to errors. For the future reference and to provide a complete information, all charts should carry a heading giving identifying information such as the name of the part, material, or equipment, number required, part number and drawing number, name of the operator, name of the analyst, method used (present or proposed), the job or process being carried out, the location (department, factory, shift), reference number, sheet number and the total number of sheets and the time and date of the study.

Example of a flow process chart

Fig. 6.6 shows a flow process chart for the manufacture of a fixed jaw (main component) of a vice. This is a worker type flow process chart.

On completion of flow process chart, the summary should be recorded in the space provided on the right hand top corner of the diagram. This consists of columns indicating present and proposed methods and various activities. The distances travelled during the process may be included in the summary. In case of present or existing method, total number of operations, inspections, transportations, delays, storages and distances travelled should be recorded. While in the case of a proposed method all data regarding present and proposed method should be recorded and in addition the information in column marked saving should be inserted under various headings.

Worker/Material/Equipment Type	*Flow Process Chart*			Sheet of
Subject : Manufacturing of Fixed Jaw	Method : Present			
Chart Beginning : Draw materials from store				
Chart Ends : Deposit Completed Vice in Store				
Charted By :	Approved By :			
Date :	Date :			
Part Name :				

Summary

Description	Present	Proposed	Saving
Operation	12		
Transport	8		
Delay	1		
Storage	—		
Inspection	1		
Total	22		

Description	Quantity	Distance (m)	Time (min)	○	⇨	D	□	▽	Remarks
To Store					●				
Draw fixed jaw costing from store				●					Plate form trolley
To shaper		30			●				
Faces on shaper				●					
Faces opposite end									
Shapes on top surface				●					
To milling M/C		20			●				
Straddle mills on bottom face				●					
Awaits for groove milling						●			
Straddle mills for grooves				●					
Examine for dimensions							●		
To lathe		10			●				
Boring on lathe				●					
Thread cuts in hole				●					
To vertical milling M/C		15			●				
Mills slot for clamping screw				●					
Mills opposite end									
Slot for clamping screw				●					
To H.T. shop		30			●				
Jaw case harden				●					
Carries to assembly shop		25			●				
Unloaded in assembly shop				●					
Carries to assembly store		10			●				

Fig. 6.6

Worker/Material/Equipment Type	*Flow Process Chart*			Sheet of
Subject : Manufacturing of Fixed Jaw Chart Beginning : Draw material from stores Chart Ends : Deposit finished Vice in Stores	Method : Present	Summary		
		Description	Present	Proposed	Saving
		Operation	12		
		Transport	7		
Charted By :	Approved By :	Delay	1		
		Storage	2		
		Inspection	1		
Date :	Date :	Total			
Part Name :					

Description	Quan-tity	Distance (m)	Time (min)	O	⇨	D	□	▽	Remarks
Fixed jaw stocked in store								●	
Draw part from store				●					
Transported to shaper		30			●				
Face on shaper				●					
Face opposite end				●					
Machined top surface on shaper				●					
Carried to milling M/C		20			●				
Milled bottom face				●					
Await for groove milling						●			
Grooves milled				●					
Examined for dimension							●		
Carried to lathe		10			●				
Bored on lathe				●					
Hole threaded				●					
Carried to vertical									
Milling machine		15			●				
Milled slot for clamping screw				●					
Milled slot opposite									
End for clamping screw				●					
To H.T. Shop		30			●				
Jaw case hardened				●					
To assembly shop		25			●				
Unloaded in assembly shop				●					
Carried to assembly store		10			●				
Being stored								●	

Fig. 6.7

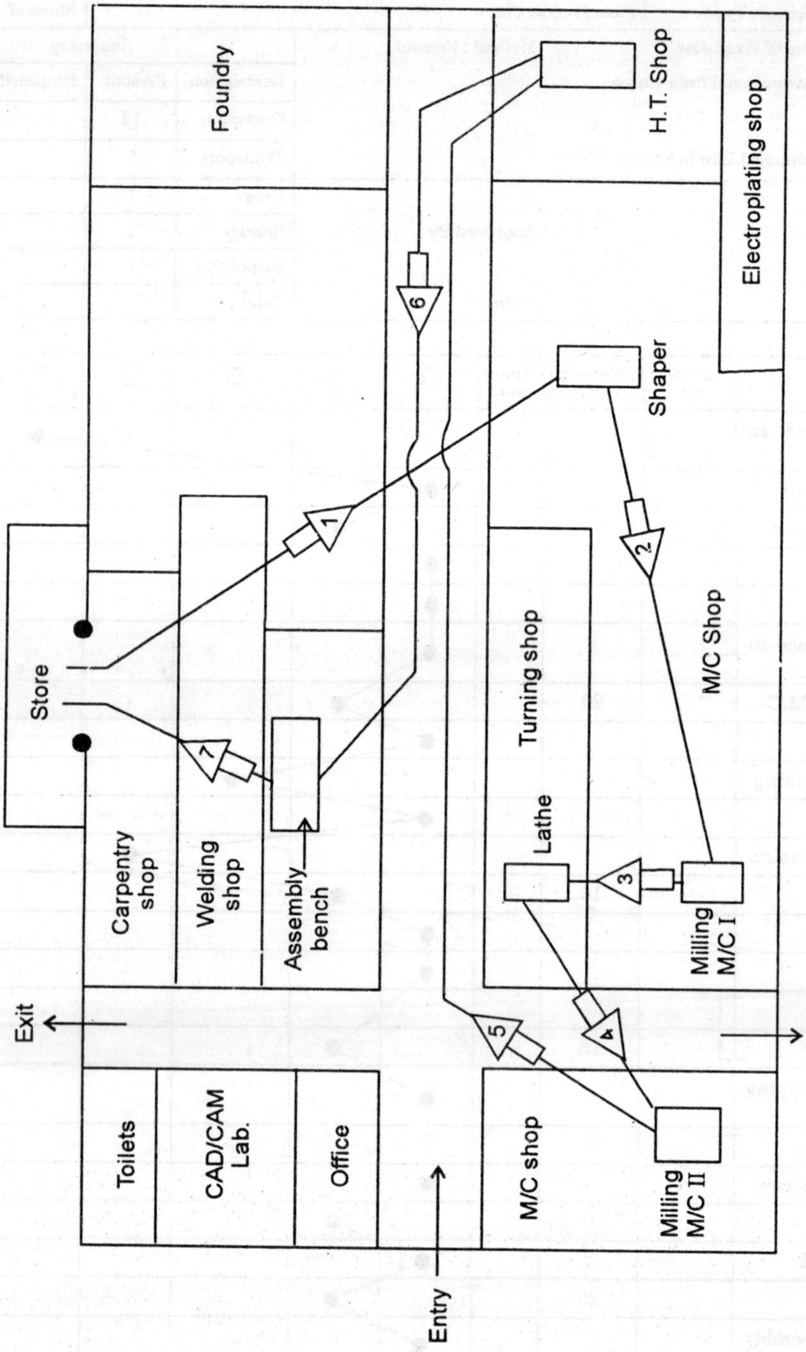

Fig. 6.8. Flow diagram : Manufacturing of fixed jaw (Present Method).

6.7. PROCEDURE FOR EXAMINE CRITICALLY

After recording the relevant facts about the job, the next step of the work study is the "Examine Critically". This is done by questioning techniques. It is the means by which the

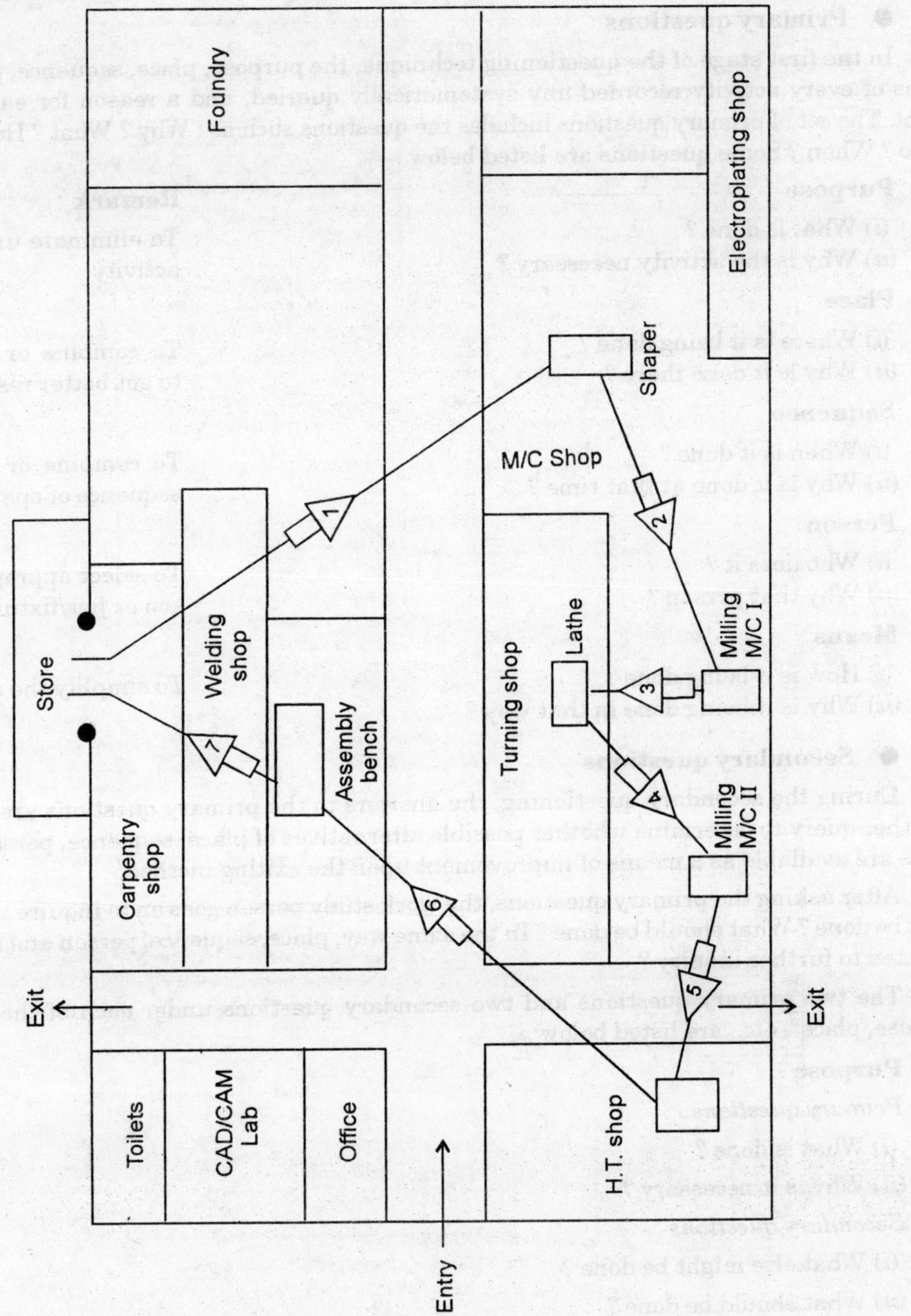

Fig. 6.9. Flow diagram : Manufacturing of fixed jaw (Proposed Method).

critical examination is conducted by asking some questions to the activities with a view to improve the work. The answers to these questions usually provide the analyst with ideas for possible improvements. The questions are asked in two stages. First stage of questioning covers primary questions and second stage of the questioning covers secondary questions.

- **Primary questions**

In the first stage of the questioning technique, the purpose, place, sequence, person and means of every activity recorded any systematically queried, and a reason for each reply is sought. The set of primary questions includes the questions such as : Why ? What ? How ? Where ? Who ? When ? Some questions are listed below :

	Remark
Purpose	
(*i*) What is done ? (*ii*) Why is the activity necessary ?	To eliminate unnecessary activity
Place	
(*i*) Where is it being done ? (*ii*) Why is it done there ?	To combine or rearrange to get better result
Sequence	
(*i*) When is it done ? (*ii*) Why is it done at that time ?	To combine or rearrange sequence of operation
Person	
(*i*) Who does it ? (*ii*) Why that person ?	To select appropriate person or jigs/fixtures
Means	
(*i*) How is it being done ? (*ii*) Why is it being done in that way ?	To simplify the operations

- **Secondary questions**

During the secondary questioning, the answers to the primary questions are subjected to further query to determine whether possible alternatives of place, sequence, persons and/or means are available as a means of improvement upon the exiting method.

After asking the primary questions, the work study person goes on to inquire : What else might be done ? What should be done ? In the same way, place, sequence, person and means are subjected to further inquiry ?

The two primary questions and two secondary questions under each of the headings "purpose, place", etc., are listed below :

Purpose

Primary questions :

(*i*) What is done ?

(*ii*) Why is it necessary ?

Secondary questions :

(*i*) What else might be done ?

(*ii*) What should be done ?

Place

Primary questions :

(*i*) Where is it being done ?

(*ii*) Why is it done there ?

Secondary Questions :

(*i*) Where else might it be done ?

(*ii*) Where should it be done ?

Sequence

Primary questions :

(*i*) When is it done ?

(*ii*) Why is it done at that time ?

Secondary questions :

(*i*) When might it be done ?

(*ii*) When should it be done ?

Person

Primary questions :

(*i*) What does it ?

(*ii*) Why does that person do it ?

Secondary questions :

(*i*) Who else might do it ?

(*ii*) Who should do it ?

Means

Primary questions :

(*i*) How is it being done ?

(*ii*) Why is it done that way ?

Secondary questions :

(*i*) How else might it be done ?

(*ii*) How should it be done ?

These questions, in the above sequence, must be asked systematically every time in a method study is under taken. They are the basis of successful method study. How they do so can be best demonstrated by directing the questions to a few of the activities in our 'Fixed jaw manufacturing' example.

Detailed examination of the flow chart (Fig. 6.7) leads to a number of questions. Let us apply the questioning technique to this chart and flow diagram (Fig. 6.8).

Q. What is done ?

Ans. Machining of the fixed jaws is carried out.

Q. Why is it done ?

Ans. Steel-castings of fixed jaws are stored in store, so for dimensional accuracy and surface finish machining is required.

Q. How is it being done ?

Ans. The jaws are collected from store and carried to different department on different machines.

Q. What else might be done ?

Ans. Departments could be arranged as per the sequence of operations. So that the distance of movement could be minimized.

Q. What should be done ?

Ans. Milling machine I and milling machine II should be placed in a machine shop nearer to each other.

Heat treatment shop should be arranged in place of small machine shop. So that the distance between milling machine II and H.T. shop will be reduced. In present method, the total distance to be moved was 140 metre. In proposed method, the total distance to be moved is 104 metre only. This way questioning technique may be applied to other activities in the flow process chart to improve the many things.

7

Movement of Workers in the Plant

7.1. INTRODUCTION

The operation process chart, flow process chart and flow diagram give an overall ideas of sequence of operations and assembly, transportation of raw material and finished or semi-finished products, delays, storages and inspections during the course of production of goods. These informations are valuable for the development of the plant layout or relayout. These charts give the information about the distance between two work stations, operation time, sequence of operation, sequence of movement location of machines and/or equipment in the plant. But this is not enough for an efficient layout of different department in the plant.

There are various activities in the plant in which workers move at irregular intervals between a number of points in the working area, with or without materials. The frequency of movement between two work station may be more. In this situation, operation process chart, flow process charts, and flow diagram will not show the frequency of movement and it is also difficult to say the exact distance between two work station when the movement is irregular. In industries movement of men or equipment or materials occurs when :

(*i*) Materials is being fed to the different machines or processes ;

(*ii*) Worker attending more than one machines ;

(*iii*) Workers are moving to stores for receiving tools and materials and delivering materials and tools to operators.

To attain maximum economy in the movements of workers and materials during the production, it is necessary to locate the various work station and machines in such a manner that the movement of the workers and materials is reduced to a minimum. The string diagram and travel charts are the useful recording technique in this direction.

7.2. THE STRING DIAGRAM

The string diagram is a one of the most useful techniques for recording and examining activities in method study. It is a scaled plan or model on which the path of workers, materials or equipment is traced by winding a string (thread) along the path of movement. The length of the path traced by man, material or equipment is measured by the length of the string (thread) used during a specified sequence of events. It is a special form of flow diagram, in which string or thread is used to measure the distance of movement between two work stations and total distance travelled by man, materials or equipment during the study. It is used when the frequencies of movement involved are large and cannot be comprehensively recorded by flow diagram. It is drawn correctly to scale. Flow diagram may be drawn approximately to scale. It is used to supplement a flow process chart. It gives the clearest picture of what is actually being done. Flow diagram simply shows the path of movement of workers, materials or equipments.

It does not show the frequencies of movement between two work stations. Flow diagram is prepared by drawing a line on scaled plants. In string diagram, movement is shown by winding a string (thread) around the pins inserted at each work station.

7.3. PROCEDURE OF CONSTRUCTING STRING DIAGRAM

- A scaled plan of the work area is first prepared on a paper and is pinned on a soft board or plywood.
- Various work stations, machines, benches, stores are marked on this plan and pins driven into it firmly at every stopping points, the heads being allowed to stand well clear of the surface (by about 10 mm), so that string can be conveniently wound around them during construction of the diagram.
- The work study person proceeds to follow the worker being investigated as he moves from point to point in doing the job. In the small working area, work study person can watch the worker without moving.
- A measured length of thread is taken and tied round the pin at the starting point of the movements. It is then led around the pins at the other points of call in the order noted on the study sheet until all the movements have been dealt with.
- The study will continue as long as the study man feels that he has got the representative picture of the journeys made by various subjects. If proper care is not exercised in deciding this the diagram will give a misleading picture and the layout based on it will be incorrect.

The thread used was measured before the study person started to make the diagram. By measuring the length remaining and subtracting this from the total length, the length used can be found. This will represent, to scale, the distance covered by the workers. It gives an overall picture of the paths of the movements of the operatives, those which are most frequently traversed being covered with the greatest number of strings, the effect being as in Fig. 7.1. If

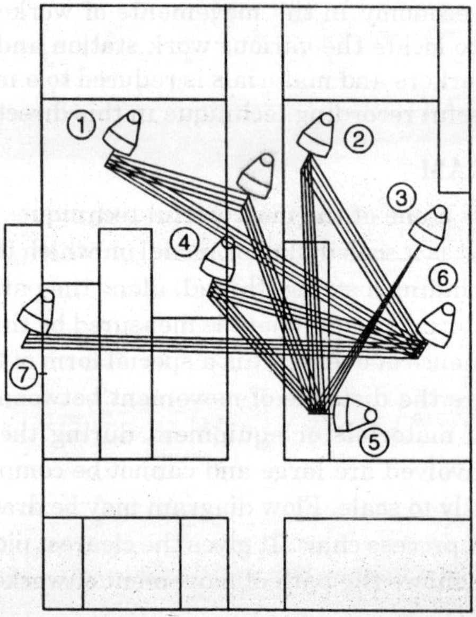

Fig. 7.1. A string diagram : 1, 2, 3, 4-Machines, 5-Tool crib, 6-Inspection bench, 7-Finished part store.

two or more workers are studied over the same working area, different coloured threads may be used to distinguish between them. Fig. 7.1 shows a string diagram for the movements of a workers attending to the needs of four workers. It is seen from the diagram that the journeys to station—5 (tool crib) and station—6 (inspection bench) are more frequent and both these stations are away from machines. If the relayout of the machines is considered and the location of tool crib and inspection bench shifted to the centre of these machines, the distance travelled by the worker could be considerably reduced ; thus arriving at a better layout.

The Examination of the diagram and the Development of the new layout can now proceed on the lines as with a flow diagram, with templates being used and the pins and templates moved around until an arrangement is found by which the same operations can be performed with a minimum movement between them. This can be ascertained by leading the thread around the pins in their new positions, starting from the same point and following the same sequence. When the thread has been led around all the points covered by the study, the length left over can again be measured. The difference in length between this and the thread left over from the original study will represent the reduction in distance travelled as a result of the improved layout. The process may have to be repeated several times until the best possible layout is achieved.

String diagram is thus a very effective means of recording and presenting the facts regarding the movements of the workers in existing layout. The string diagram is a useful aid in explaining proposed changes to management, supervisors and workers. The string diagram, however have limitations. String diagram do take a long time to construct, however, and when a great movements along complex paths are involved the diagram may end up looking like a forbidding maze of criss-crossing lines. It is difficult to draw a fair conclusion from such a diagram. To overcome this difficulty the "travel chart" is prepared. The travel chart is a quicker and more manageable recording technique.

7.4. THE TRAVEL CHART

The travel chart is a two-way matrix table for presenting quantitative data about the movements of workers, materials or equipment between any number of work stations over any given period of time. The travel chart is always a square, having within it smaller squares. Each small square represents a work station. The main purpose of the travel chart is to make an efficient layout of plant and shops so that the distance travelled by man and materials during the manufacture of a product is minimum.

Fig. 7.2 shows a typical travel chart. It records the movements of a workers moving to the various work stations or machines in the plant.

The squares from left to right along the top of the chart represent the work station 'from where' movement takes place ; those down the left hand edge represent the work station 'to which' the movement is made. For example, consider a movement from station 2 to station 8. To record this, the study man enters the travel chart at the square numbered 2 along the top of the chart, runs a pencil down vertically through all the squares underneath this one until he reaches the square which is horizontally opposite the station marked 8 on the left-hand edge. This is the terminal square, and the study man will make a mark in that square to indicate one journey from station 2 to station 8. All journeys are recorded in the same way, always starting at the top, travelling vertically downwards, and ending in the square opposite the work station of arrival, as lead from the left-hand edge. The study man just places a small tick mark or other mark in the terminal square to record the journey.

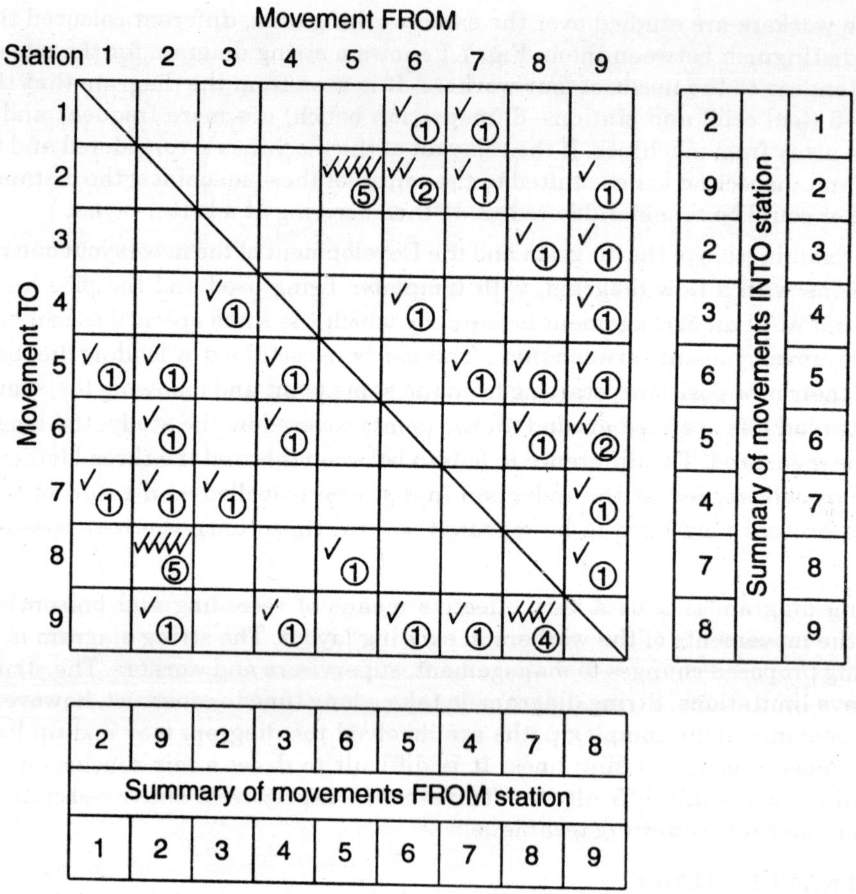

Fig. 7.2. Travel chart.

7.5. WEIGHTED TRAVEL CHART

The weighted travel chart is the travel chart on which the movement is shown with variation in volume or weight of the materials moved. This records movements over a given period of time (normally one shift or one day). The recording journeys made between locations shows the volume or weight of the materials. At the end of the time period, the total traffic weight or volume between each set of locations is entered on the tubular chart and used as the basis of examination.

An example of a weighted travel chart is shown in Fig. 7.3. In this example a company which undertake to manufacture ten items in their programme. The management desires a planned layout of various shops for the efficient manufacture of these product. The different shops such as *a*, *b*, *c* ... *i* shows at the top of the chart, from which materials to be moved 'INTO' the shops which are shown at the left side of the chart vertically. As per the simple travel chart, here also movements are indicated From-To by a tic. Along with tick mark, the unit load per trip is also written near the each mark. For example, from *f* to *a* first mark shows 40 unit load trips, second mark shows 10 unit load trips. From *g* to *a*, 40 unit load trips are shown. In summary 'INTO', number of trips indicated 3 and total unit load trips 90. It is seen from the chart that maximum number of trips are 220 from '*b*'. From chart it is seen that maximum

Movement-FROM (columns) / Movement-TO (rows)

Movement-TO \ Movement-FROM	a	b	c	d	e	f	g	h	i	Summary (INTO): No. of trips	Summary (INTO): Total unit load trip
a						√40 √10	√40			3	90
b			√10				√20	√20		3	50
c				√40		√40				2	80
d						√ –	√10			2	10
e		√100 √50								2	150
f	√20								√30 √30 √40	4	120
g									√ – √ – √40	3	40
h									√10	1	10
i	√ – √30	√40 √30	√10	√30	√40 √30					8	210
Summary (FROM): No. of trips	3	4	2	2	2	4	3	1	7		
Summary (FROM): Total unit loads trips	50	220	20	70	70	90	70	20	150		

Summary of movements FROM station

Fig. 7.3. Weighted travel chart.

number of unit loads trips are from *b* to *e* and *b* to *i*. Similarly maximum unit load trips are 210 '*INTO*' *i*. From the chart it is seen that these are FROM '*b*' to *i* and *e* to *i*. Thus, an improved plant layout can now be prepared on the basis of this logic and the information gathered from the travel chart.

7.6. MULTIPLE ACTIVITY CHART

The multiple activity chart is used to record on one chart the activities of one subject in relation to another. It is a graphical representation of the co-ordinated activities of more than one subject. It records on a common time scale, the activities of more than one subject (worker, machine or equipment). It records the working and waiting time of two or more men, machines or any combination of men and machines. The chart shows very clearly periods of idleness on the part of any of the subjects during the process. A study of the chart often makes it possible to rearrange these activities so that such ineffective time is reduced to a minimum.

A multiple activity chart is an extremely useful tool in planning a production and maintenance work. When expensive plant cannot be allowed to remain idle longer than is absolutely necessary. It can also be used efficiently in organising teams of operatives on mass-production work. It can also be used to determine the number of machines or equipments which an operative or operatives should be able to attend. This chart is used to obtain higher utilization factor of machines or/and men involved in multiple activities. It is used to determine the most economical patterns of men and machines relationship, taking into consideration the production requirement and cost relationship.

7.7. MAN AND MACHINE PROCESS CHART

Man and machine process chart is a special type of multiple activity chart in which the activities of the different operatives and machines are recorded in terms of working time and idle time. With the advancement in the machine manufacture, semi-automatic and automatic machines are increasingly employed in manufacturing processes. This introduce idle operator time for a portion of cycle during which the machine is operating on automatic feed. This information is obtained from man-machine chart in a graphic form. The analysis of the chart may indicate possibility of an operator handling two or more machines, thus improving production efficiency. The times may be recorded by ordinary wristwatch, by stop-watch or by electronic timing, according to the duration of the various periods of work and idleness.

In constructing the man-machine process chart, the first step is to record all information regarding the job ; namely the description of operation being charted, the machine used, part number and drawing number, the present method, special tools, or equipment used, the names of the analyst and the operator and the date and time when the study was undertaken. These informations are written at the top of the chart which is usually called heading of the chart.

The second step is to define activities involved during the operation and to indicate it on chart. The activities are usually classified in three categories :

(1) **Independent activities.** When an operator is inspecting a machined part or when he is working independently, it is called independent activity of man. Similarly when the machine is operating on automatic feed during an operation, the activity of machine is called independent activity of machine.

(2) **Combined activities.** When man and machine both are involved in working on a job giving hand feed, servicing machine, loading and unloading of job from machine, the activities are called combined activities.

(3) **Idle man/machine.** When man is waiting for the machine to complete the automatic cut or when he is waiting for other operators to complete their work, the man is called idle. Similarly when man is doing independent work such as inspection of a machined component or he is tending other machine, the machine is called idle during that period. The activities are shown by following symbols on chart :

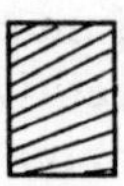 Independent activities of men and machines.

 Combined activities of men and machines.

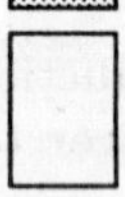 Idle time of men and machines.

MAN AND MACHINE PROCESS CHART

Subject Charted : Machining Component   Method : Present

Drawing No. : ____ Part No. ____ Chart No. ____ Sheet ____ of ____

Chart Begins : Loading on m/c   Charted by ____

Chart Ends : Inspection   Date : ____

Min.	MAN Description	Time	MACHINE Description	Time
0–2	Loading of the job	2	Being loaded	2
2–10	Idle	8	Machining	8
10–13	Unloading of the job	3	Being Unloaded	3
13–18	Inspection	5	Idle	5

SUMMARY

	Man	M/C
Idle time (min.)	8	5
Work time (min.)	10	13
Cycle time (min.)	18	18
% utilization	55.55	72.22

Fig. 7.4. Man and machine process chart (Present method).

The use of the man and machine process chart can best be shown by an example.

Fig. 7.4 shows the one cycle of a man and machine process chart for the existing method of working on an operation involving the following activities :

- Stopping the machine and unloading—3 min
- Machining time (Automatic feed)—8 min
- Loading the job on machine—2 min
- Inspection of the completed job—5 min

It is observed from the Fig. 7.4 that the operator remain idle for almost 55.53% of cycle time. By rearranging the sequence of operations, it is possible to reduce the cycle time. The inspection of machined component can be carried out—when the machine is switched on automatic operation. However, when operator idle time is on the higher side, it is a better practice to examine the possibilities of operator tending more than one machine.

Fig. 7.5 shows the activities of one man tending two machines. It is seen from the chart that nothing substantial has been gained by the operator tending two machines. In fact, in this particular case, utilization percentage of man increases to 76.92 while utilization percentage of m/c I and II decreases to 50%. The economical production of the component depends on the hourly rate of the man and the machine. From this percentage utilization of man and machine, it is difficult to say that which one is economical. A final decision should be made only after comparing the manufacturing cost of the component in éach case. In the above case, it is possible to reduce the cycle time by rearranging the sequence of operations. The Fig. 7.6 shows the rearranged sequence of operation. The utilization factor of man as well as machines is improved by change in sequence of operation. It is seen that the man now idle for only one minute and each machine is idle for eight minutes. Though the final decision regarding the man-machine combination is not complete yet. The choice of one operator tending one machine or more machines can now be decided on the operating cost of the machines and the wages earned by the operator. This is explained as follows :

If the machining cost is, say Rs. 60 per hour and the wages earned by the operator is Rs. 27 per hour, then

$$\text{Cost of operating machine for 1 minute} = \frac{60 \times 100}{60}$$

$$= 100.00 \text{ paise per min} = \mathit{1.00\ Rs.\ per\ min.}$$

$$\text{Cost of operator working for 1 minute} = \frac{27 \times 100}{60} = \mathbf{45\ paise/min.}$$

Case I. One operator tending one machine :

The cycle time is 18 minutes. Hence cost of production of component is 18 (machining cost and operator cost)

i.e. 18 (1 × 100 + 1 × 45) = 18(145) Rs. = 26.10 per component.

Case II. One operator tending two machines.

In this cycle time is 26 minutes (Fig. 7.5) and each machine produces one component. The time required to produce one component is there for 13 minutes and the cost is, 13 (2 × 100 + 1 × 45) = 3185 paise or Rs. 31.85 per component. This is higher than the one operator tending one machine.

MAN AND MACHINE PROCESS CHART

Subject Charterd : Machining Component Method : Present/ ______

Drawing No. : ____ Part No. ____ Chart No. ____ Sheet ____ of ____

Chart Begins : ________ Charted by ____

Chart Ends : ________ Date : ____

MAN		Machine I		Machine II	
Description	Time (min.)	Description	Time (min.)	Description	Time (min.)
Loading M/C I	2	Being loaded	2	Idle	2
Loading M/C II	2	Machining	8	Being loaded	2
Idle	6			Machining	8
Unloading M/C I	3	Being Unloaded	3	Idle	6
Inspection of Component on M/C I	5	Idle	13	Being Unloaded	3
Unloading M/C II	3			Idle	5
Inspection of Component on M/C II	5				

Time scale (min.): 0 – 26

SUMMARY

	Man	M/C I	M/C II
Idle time (min.)	6	13	13
Work time (min.)	20	13	13
Cycle time (min.)	26	26	26
% utilization	76.92	50	50

Fig. 7.5. Man and machine process chart.

MAN AND MACHINE PROCESS CHART

Subject Charted : Machining Component   Method : Proposed

Drawing No. : _____ Part No. _____ Chart No. _____ Sheet _____ of _____

Chart Begins : __________   Charted by _____

Chart Ends : __________   Date : _____

MAN		Machine I		Machine II	
Description	Time (min.)	Description	Time (min.)	Description	Time (min.)
Loading M/C I	2	Being loaded	2	Machining	2
Unloading machine II	3	Machining	8	Being unloaded	3
Inspection on machine II component	5			Idle	8
Unloading M/C I	3	Being Unloaded	3		
Loading M/C II	2	Idle	8	Being loaded	2
Inspection on machine I component	5			Machining	6
Idle	1				

SUMMARY			
	Man	M/C I	M/C II
Work time (min.)	20	13	13
Idle time (min.)	1	8	8
Cycle time (min.)	21	21	21
% utilization	95.24	61.90	61.90

Fig. 7.6. Man and machine process chart (Improved).

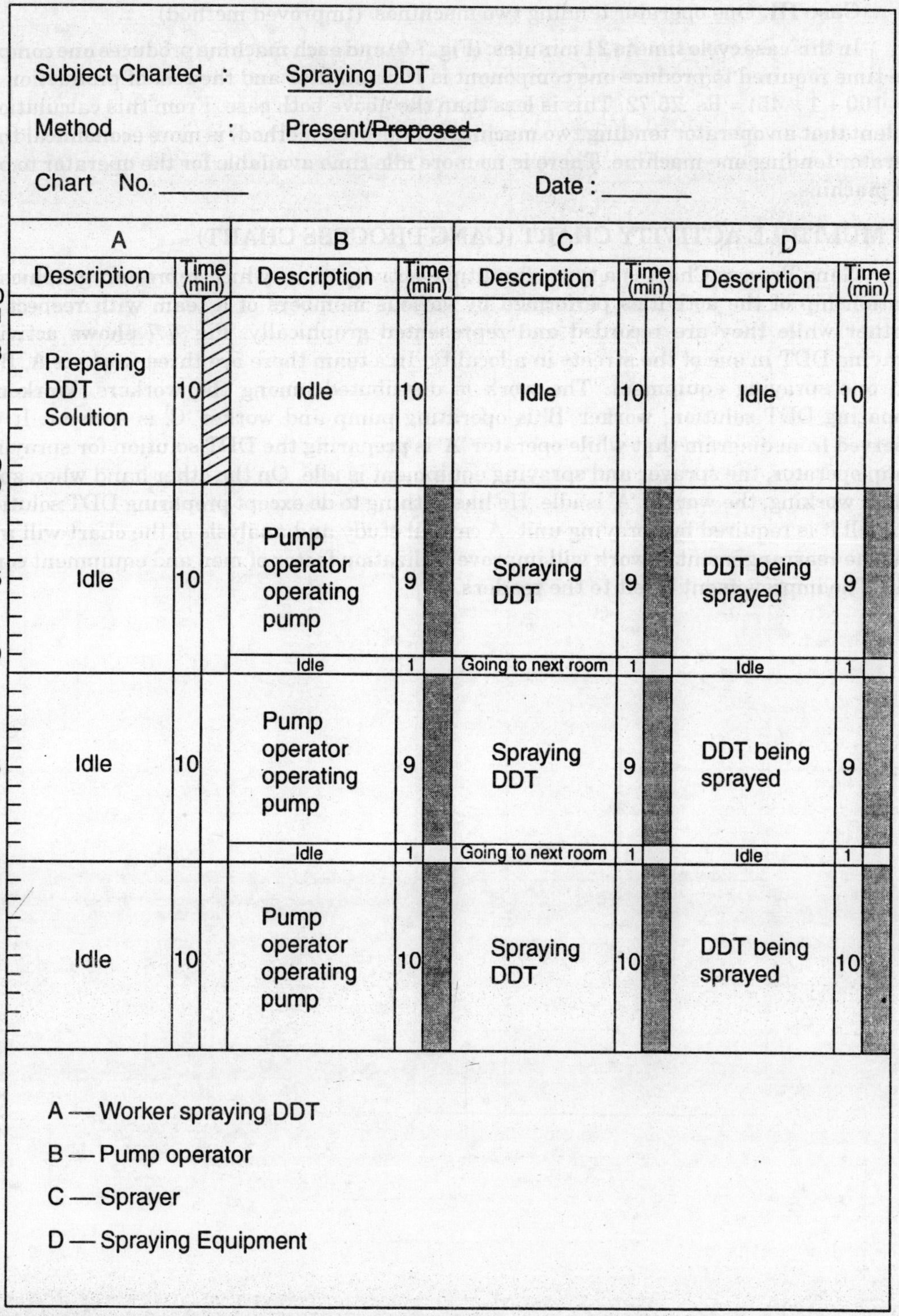

Fig. 7.7. Multiple activity chart (Gang chart).

Case III. One operator tending two machines. (Improved method)

In this case cycle time is 21 minutes. (Fig. 7.6) and each machine produces one component. The time required to produce one component is 10.5 minutes and the cost of production is 10.5 (2 × 100 + 1 × 45) = Rs. 25.72. This is less than the above both case. From this calculation, it is evident that an operator tending two machines (improved method) is more economical than the operator tending one machine. There is no more idle time available for the operator to operate 3rd machine.

7.8. MULTIPLE ACTIVITY CHART (GANG PROCESS CHART)

Gang Process Chart is a type of multiple activity chart which represent graphically the relationship of the activities performed by various members of a team with respect to one another while they are recorded and represented graphically. Fig. 7.7 shows activities of spraying DDT in one of the streets in a locality. In a team there are three workers A, B and C and one spraying equipment. The work is distributed among the workers. Worker 'A' is preparing DDT solution, worker 'B' is operating pump and worker 'C' is sprayer. It will be observed from diagram that while operator 'A' is preparing the DDT solution for spraying, the pump operator, the sprayer and spraying equipment is idle. On the other hand when spraying unit is working, the worker 'A' is idle. He has nothing to do except preparing DDT solution and wait till it is required by spraying unit. A careful study and analysis of the chart will indicate that the rearrangement of work will improve utilization factor of men and equipment considerably. The improvement is left to the readers.

8

Movements at the Work Place

8.1. INTRODUCTION

We have examined procedures of a general nature for improving the effectiveness with which complete sequences of operations are performed and with which material flows through the working area. We have discussed in previous chapter, the methods of movements of workers around the working area and the relationships between men and machines or workers working together in groups.

In this chapter we look at one worker working at a work place, bench or table. By the application of principles of motion economy it is possible to eliminate unwanted movements.

8.2. THE PRINCIPLES OF MOTION ECONOMY

These principles have been developed by Frank and Lillian Gilbreth in 1923. Subsequently then have been elaborated by Mogensen, Ralph Barnes, David Porter and other research workers. There are a number of "Principles" concerning the economy of movements of human body. These principles have been developed from experience, which form a basis for the development of improved methods at the work place. In recent days of industrial development and mass production of various commodities, thousands of physical motions are repeated by workers in course of performing their routine work. It is essential to determine the correct and efficient motion pattern to perform a given task. By the application of these principles it is possible to improve the existing methods of working to achieve higher efficiency. These principles may be grouped under three categories :

(1) Use of human body

(2) Arrangement of work place

(3) Design of tools and equipments.

They do form a very good basis for improvement in the efficiency and reducing the fatigue of manual work. These ideas are described here in a somewhat simplified manner as under :

(1) Use of human body. Following are the rules to be observed for motion economy related to the human body :

- Both hands should begin and complete their movements at the same time.
- The two hands should not be idle at the same time except during periods of rest. This is required to balance the movements of hands as far as possible. Human body is symmetrical about the vertical axis. When only one hand is working and the other is idle, maximum efficiency is not obtained. It is also tiring to keep one hand motionless while the other is working.
- Motions of the arms should be symmetrical, simultaneous and in opposite directions as shown in Fig. 8.1 (*a*). However, when parts are to be moved toward or away from operator, both the hands should move in the same direction as shown in Fig. 8.1 (*b*).

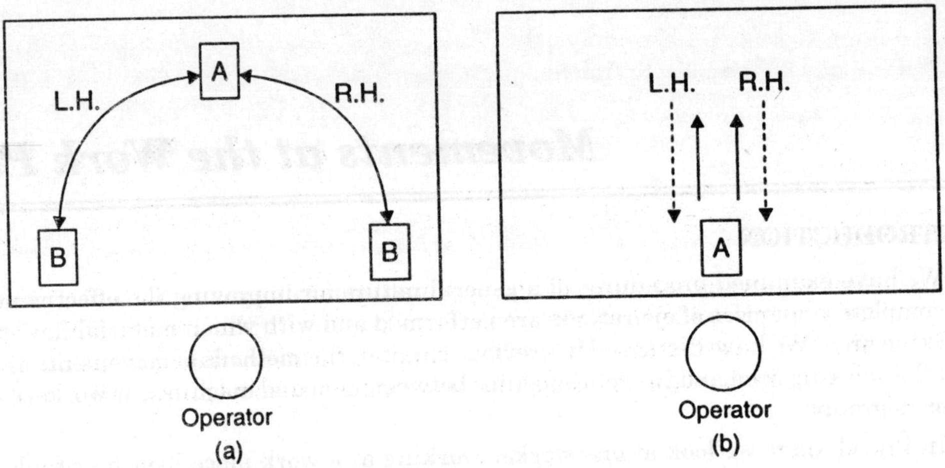

Fig. 8.1. Simultaneous motions of hands.

- **Hands and body motions should be confined at the lowest classification at which it is possible to do the work satisfactorily. This will achieve maximum performance and minimum fatigue. This classification is built up on the pivots around which the body members must move. Various hand and body motions are classified in the following five categories. They are arranged in order such that the first one give least fatigue and is most economical.**

  (1) Finger motions.

  (2) Finger and hand (wrist) motions.

  (3) Finger, hand and fore arm motions.

  (4) Finger, hand, fore arm and upper arm motions.

  (5) Finger, hand, fore arm, upper arm and Torso (body) motions.

  It may be seen from the above classification that only the fifth class of motions require a change of position while first four do not require any change in posture. The attempts should always be made to restrict movements to the lowest classification as far as practicable.

- **As far as practicable, momentum should be employed to help the worker but should be reduced to a minimum whenever it has to be overcome by muscular effort.**
- **Attempt should be made to use continuous curved movements in preference to straight line motions involving sudden and sharp changes in direction.**
- **"Bellastic" (free-swinging) movements should be preferred as they are faster, easier and more accurate than restricted or controlled movements.**
- **The work should be arranged on a table so that the eye movements are easier and confined to a comfortable area, without the need for frequent changes of focus. Worker has not to turn his head/neck to locate a part or tool while working on a table.**

(2) Arrangement of the Work place. Following are the rules to be observed in designing a work place layout to obtain desired motion economy.

- **Definite and fixed position should be provided for all tools and materials while designing work stations. This permit operators in habit formation, thus, reducing fatigue and ineffective time in searching them.**

- Tools and materials should be pre-positioned to reduce searching.
- To deliver the materials for assembly operations, gravity feed, bins and containers should be used. Similarly for disposal of finished parts also, bins or containers should be used.
- The work station should be positioned in front of the operator (area A) and the tools and the materials should be located within the maximum working area and as near to the worker as possible (Fig. 8.2).

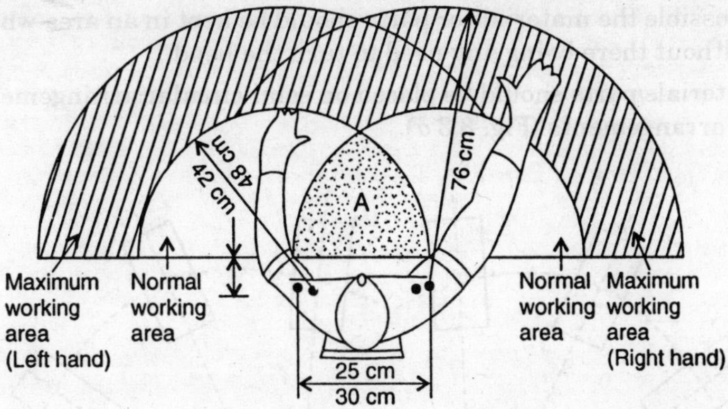

Fig. 8.2. Normal and maximum.

- The arrangement of materials and tools should be such that it will permit the best sequence of motions while performing a job.
- The hands should be left free to do useful work. This is possible by using "Drop deliveries" or "Ejectors" to dispose off finished goods wherever possible.
- Provision should be made for adequate lighting, appropriate height of the work place and the height of the seat to permit good posture while working. The height of the work place and seat should be arranged to allow alternate standing and sitting.
- To reduce eye fatigue, the colour of the work place should contrast with that of the work.

(3) **Design of Tools and Equipments.** Following are the rules to be observed in designing tools and equipments to obtain desired motion economy.

- The hands should be kept free as far as possible. Particularly the hands should be relieved of all work of "holding" the workpiece. This can be done by using Jigs, fixture or foot-operated device.
- Two or more tools should be combined wherever possible. The design of the combined tools should be such that each constituent of the combination is used in succession during operation.
- When the work is done by fingers only, the load should be distributed in accordance with the inherent capacities of the fingers. The thumb, middle finger and index finger have more capacity than the other two fingers. In type-writing all fingers are used simultaneously and performs some specific movements.
- Handles on cranks and screw drivers should be so designed that as much of the surface of the hand as possible can come into contact with the handle.
- Levers, crossbars and hand wheels should be positioned at such place that the operative can use them with the least change in body position and the greatest "mechanical advantage".

Fig. 8.2 shows the normal working area and the maximum working area on the work bench for the average operative. The normal working area involves the movements of finger, wrist, and elbow. While maximum working area involves the movements of shoulder also.

8.3. SUMMARY ON WORKPLACE LAYOUT AND SIMPLIFICATION OF MOVEMENTS

- For the similar work of each hand, there should be a separate bin of materials or parts for each hand.
- As far as possible the materials or parts should be kept in an area where the eyes can locate it without there being any need to turn the head.
- Bins of materials/parts should be placed on semi-circular arrangement in preference of circular arrangements (Fig. 8.3 *a*).

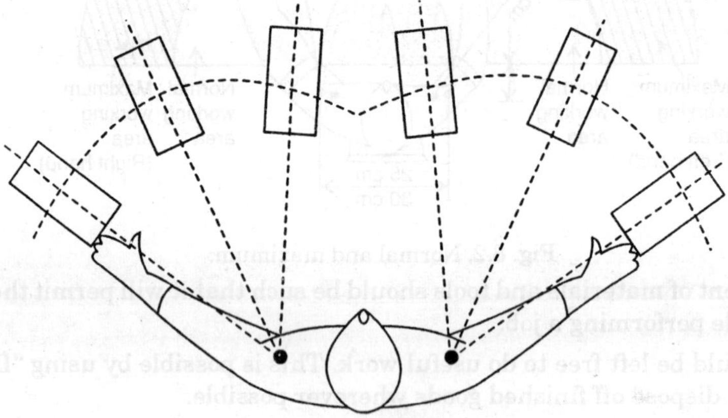

(a) Semi-Circular Arrangement

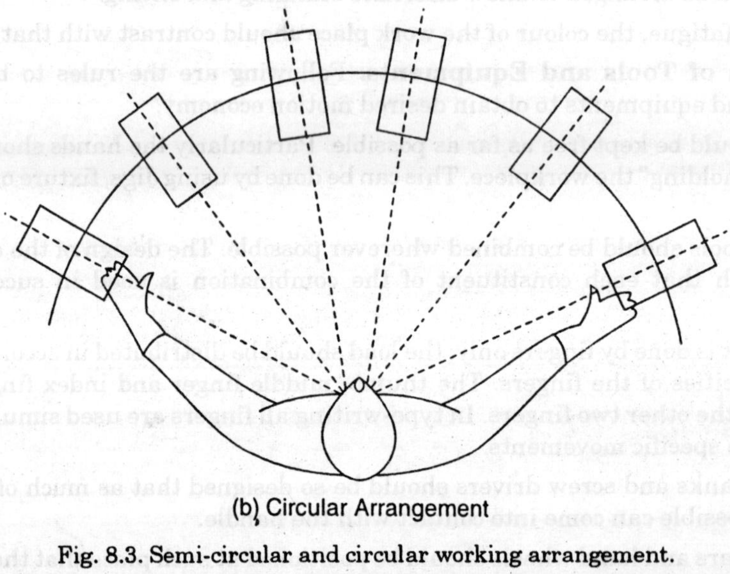

(b) Circular Arrangement

Fig. 8.3. Semi-circular and circular working arrangement.

- Use ergonomic principles in designing the work place. In a sitting position a following dimensions are recommended.
  - O Work surface height should be around 65—72 cm.
  - O Work surface thickness should be 5 cm maximum.
  - O Leg clearance should be 40 cm at knee level and 60 cm for the feet.
  - O Seat height should be 36—45 cm (adjustable to each worker).
- Use various bins and motion economy device to suit the nature and shape of the material which influence its position in the layout.
- Rhythm and symmetry of movements should be maintained in using hand tools. The operator should be able to pickup or put down a tool with the least possible disturbance to the rhythm and symmetry.
- Tools should be easy to pickup and replace. They should have an automatic return.
- Finished work should be dropped down a hole or a chute.

  If the operation is an intermediate one, finished goods should be placed in a container in such a way that the next operative can pick it up easily.

  Finished work may be put in a container placed so that hand movements are kept to a minimum.
- Wherever possible use pedals or knee-operated levers for locking or indexing devices on fixtures or devices for disposing of finished work.
- Jigs and fixtures also should be designed in such a way that the principles of motion economy is justified.

8.4. THE TWO HANDED PROCESS CHART

The two handed process chart is also known as operator process chart or left hand and right hand process chart. In this chart the movements of an operator's hands are recorded. It shows the relationship of two hands with respect to one another. This chart is a specialized form of process chart. It shows the movement of two hands of the operative in relation to one another. The activities of two hands are recorded on a time scale.

Two handed process chart is usually prepared for repetitive, manual type of operation such as assembly of radio components, watches, T.V. components, clocks, etc. In this chart, the operations are recorded in more detail than is done on flow process chart. The information recorded on this chart is analysed and by application of principles of motion economy, inefficient and unnecessary movements are eliminated ; and the work place is designed for better efficiency. The two handed process chart employs the same symbols as the other process charts ; however the symbols are accorded slightly different meanings :

O Operation is used for the activities of grasp, position, use, release, etc. of a tool, component or material.

⇨ Transport represents the movement of the hand or limb to or from the work, a tool or material.

D Delay indicates the time during which the hand or limb being charted is idle.

∇ Hold is used in place of storage. The 'storage' symbol is redesignated as hold and is used to represent the activity of holding the work, a tool or material.

The 'inspection' symbol is not used in this chart because the hand movements when an operator is inspecting an article may be classified as "operations". Here "operation" activity includes "inspection" also. During 'operation' it is also checked that the article functioning properly. So that there is no need of any 'inspection' separately.

8.5. EXAMPLE OF A TWO HANDED PROCESS CHART

Fig. 8.4 shows the assembly of a handle. It consist of three component, namely plate, handle and screw. The components are small and can be conveniently handled by either hand. The work is confined to a single work station. Fig. 8.6 is the two handed process chart prepared for the assembly shown in Fig. 8.4.

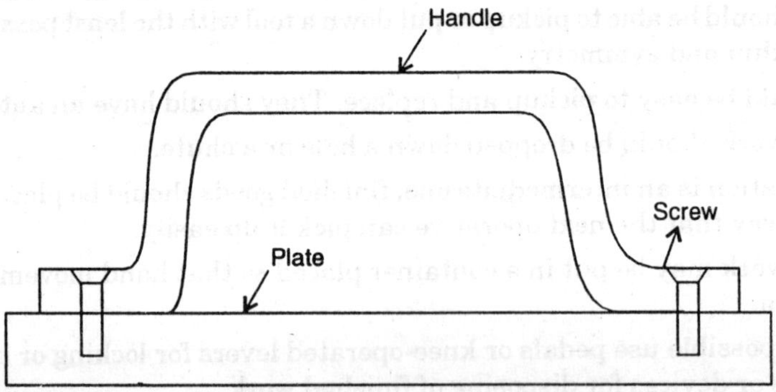

Fig. 8.4. Handle Assembly.

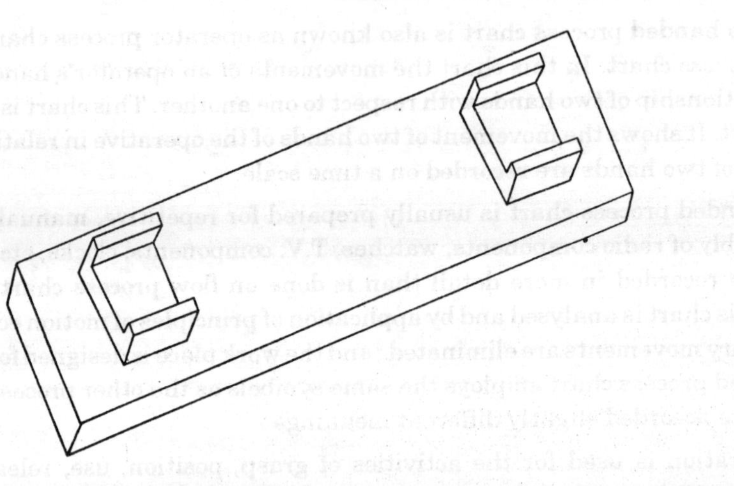

Fig. 8.5. Fixture for Handle Assembly.

Upon examining the completed operator process chart, the analyst is likely to note that two things call for corrective action. The first is that for a relatively large proportion of the work cycle, the operator's left hand holding the plate and handle while the right hand is getting the screws, and the second is that the bins arrangement is such that it is inconvenient for the operator to reach to the bin.

OPERATOR PROCESS CHART

Chart No. ______ Sheet No. ______
Drg. No. ______ Part No. ______
Location : Assembly Section
Method : Present Operator : ABN

SUMMARY

	Present		Proposed		Saving	
	L.H.	R.H.	L.H.	R.H.	L.H.	R.H.
Operation	2	4				
Transportation	3	4				
Hold	4	—				
Delay	1	2				
Time (min.)	0.35					

Layout :

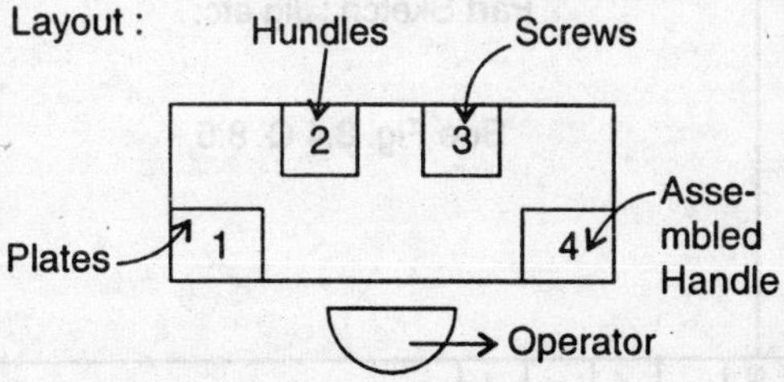

Part Sketch ; Jig etc.

See Fig. 8.4

Left hand Description	○	⇨	▽	D	Time	○	⇨	▽	D	Right hand Description
To Plate in Bin 1		•					•			To handle in Bin 2
Pick up plate	•					•				Pick up handle
To work position		•					•			To plate at work position
Delay				•		•				Position handle over plate
Hold			•				•			To screw in bin 3
Hold			•			•				Pick up screw
Hold			•				•			To handle
Hold			•			•				Fasten handle with plate (screw handle)
To bin 4		•							•	Idle
Drop assembly in bin 4	•								•	Idle

Fig. 8.6. Work station layout and operator process chart for the assembly of a handle (Present method).

OPERATOR PROCESS CHART

Chart No. ______ Sheet No. ______

Drg. No. ______ Part No. ______

Location : Assembly Section

Method : Proposed Operator : ABN

SUMMARY

	Present L.H.	Present R.H.	Proposed L.H.	Proposed R.H.	Saving L.H.	Saving R.H.
Operation	2	4	7	7		
Transportation	3	4	7	7		
Hold	4	—	—	—		
Delay	1	2	—	—		
Time (min.)	0.35		0.25		0.10	

Layout :

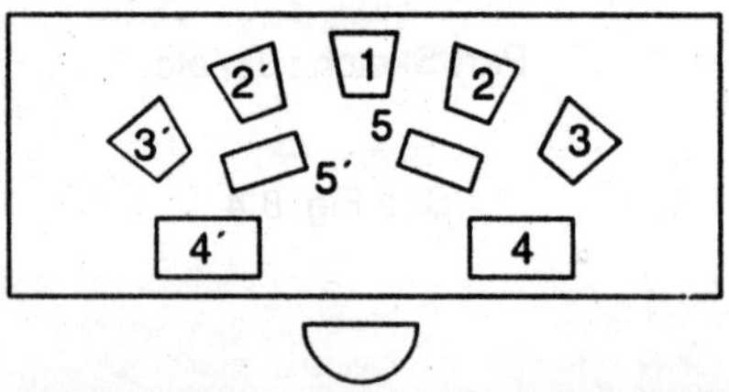

Part Sketch ; Jig etc.

See Fig. 8.4 Q. 8.5

Left hand Description	○	⇨	▽	D	Time	○	⇨	▽	D	Right hand Description
To plate in bin 1		•					•			To plate in bin 1
Pick up plate	•					•				Pick up plate
To fixture 4′		•					•			To fixture 4
Position in fixture	•					•				Position in fixture
To handle in bin 2′		•					•			To handle in bin 2
Pick up handle	•					•				Pick up handle
To fixture 4′		•					•			To fixture 4
Position over plate	•					•				Position over plate
To screw in bin 3′		•					•			To screw in bin 3
Pick up screw	•					•				Pick up screw
To fixture		•					•			To fixture
Screw in handle	•					•				Screw in handle
To chutehole 5′		•					•			To chute hole 5
Drop assembly in chute	•					•				Drop assembly in chute

Fig. 8.7. Work station layout and operator process chart for the assembly of a handle by an improved method.

With regard to the holding activity, the principles of motion economy suggest the use of a fixture of some type. In this case, the fixture of the kind shown in Fig. 8.5 is used. It consists of a iron plate with a depressed block. The bottom plate may be positioned between these two blocks. So that both hands will remain free. Two fixtures are used so that two plates can be positioned by L-H and R-H, and two assembly of the handles are obtained at a time.

From the sketch of the revised layout of the work place, it can be seen that other improvements have been made on the basis of other principles of motion economy (Fig. 8.7).

The bins of suitable sizes containing the handles and screws are duplicated and arranged in such way that both hands can work simultaneously. Furthermore as shown in Fig. 8.7 they are placed radially making it possible to be arranged within normal working area, thus reducing movement of both hands.

In the improved method two holes are provided on the table in front of the fixture in which the assembled door handles are dropped so that no time is lost for starting the next cycle.

A chair of suitable design and height for comfortable working is provided. A provision of backrest is made on the chair to support the lower part of the spine. The provision of these facilities does not tire out the operator and the ultimate output is increased considerably.

8.6. MICRO MOTION STUDY

Micro motion study is a recording and analysing technique of time of elements which are of very short duration and repeated many times. Its purpose is to find the best method of working. Both film and video cameras can be used to record activities at the work place. The record can be obtained by taking a cine film of the operation at normal speed of 16 frames/second or by using a constant speed camera exposing film at 1000 frames/min. Although film provides greater flexibility in terms of filming and play back speeds, video is generally cheaper and easier to use. With a good quality video playback unit, it is possible to have light-quality slow motion and still frame facility. The advantages of film and video over direct observation are that

- the film obtained serves as a permanent record of the work studied and can be re-examined at a later date, if needed ;
- if permit greater detailing than eye observation ;
- it can be projected at any time and can be shown to number of operators during training period ;
- it provide a positive record ;
- they help in the development of the work study persons themselves ;
- the operation can be studied better by projecting it at any desired speed ;
- the film can be run backwards and reprojected to locate clumsy and awkward movements which are subsequently rectified or eliminated.

Micromotion study is expensive compared to other methods of work simplification. It should, therefore, be undertaken only when other methods of work simplification are thoroughly applied and examined and when its use can be economically justified. It requires costly equipments and apparatus together with high degree of skill on the part of work study person in taking and analysing the film. Its use, therefore, should be restricted to highly repetitive operations which involve many operators and are likely to continue for a considerable period.

TABLE : 8.1

Therbligs and their Symbols

Therbligs	Symbol	Letter Abbreviation	Colour
Search		Sh	Black
Find		F	Gray
Select	→	St	Light Gray
Grasp	∩	G	Lake red
Transport loaded	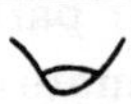	TL	Green
Position	9	P	Blue
Assemble	#	A	Violet
Use	U	U	Purple
Disassemble	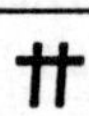	DA	Light violet
Inspect		I	Burnt ochre
Preposition		PP	Sky Blue
Release load		RL	Carmine red
Transport empty		TE	Olive green
Rest for overcoming fatigue	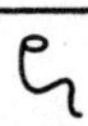	R	Orange
Unavoidable delay		UD	Yellow ochre
Avoidable delay		AD	Lamon yellow
Plan	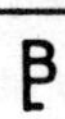	Pn	Brown
Hold		H	Gold ochre

There is another process chart which sometimes used by an organization to describe an existing motion pattern. This is a simultaneous motion cycle, or SIMO chart. This has now been replaced by the use of predetermined time standards charts (PTS charts). In general, SIMO chart describes the co-ordinated activities of an operator's body members in terms of 18 fundamental motions, which are known as therbligs.

Gilbreth observed that in majority of manual operations performed, there are certain common movements. He studied various motion patterns in great details and concluded that the immense varieties of manual work performed in factories and elsewhere can be sub-divided into 18 fundamental movements which is called 'Therblig' (Gilbreth spelt in reverse order). Each therblig has a specific colour, symbol and letter for recording purpose. This is shown in Table 8.1.

8.7. SIMULTANEOUS MOTION CHART

SIMO chart, records simultaneously on a common time scale, the activities of two hands while performing an operation under investigation. The observations are recorded in therbligs and time in wink (1/2000 min.). Such a record is usually obtained from frame to frame analysis of film taken during the performance of the operation.

In construction of a Simo-chart aline film is taken when the operation is in progress. This film is subsequently developed, projected and analysed. This needs expensive photographic and projecting equipment and skill in photographic technique. This condition restricts the use of Simo-chart only to operations which are highly repetitive in nature and are carried out by many operators for a considerable time.

With the help of these charts the ineffective and inefficient movements and delays can be found out quickly and it is possible to eliminate ineffective movements, rearrange sequence of operation of working and obtain better work distribution.

The micro motion study film is used in the construction of the SIMO chart.

The film is projected on a screen and analyst observe it several times and get acquainted himself with various therbligs involved in the operation. After this he determines time required for various therbligs involved in the operation. This time is recorded in the operation. This time is recorded on transfer sheets. Transfer sheet is an intermediate step between the analysis of film and preparation of SIMO chart. The transfer sheet form is shown in Fig. 8.8. From the data recorded on transfer sheet, SIMO chart is prepared. The Fig. 8.9 shows the film of SIMO-chart. The chart is then analysed and unnecessary movements of left hand and right hand is eliminated.

MICROMOTION TRANSFER SHEET

Operation : Department : Operator :	Film No. : Method : Present/Proposed Date :	Sheet No. : Analyst :

Clock Reading	Subtracted Time (Wink)	Therblig symbol	L.H. Description	Clock Reading	Subtracted Time (Wink)	Therblig Symbol	R.H. Description

Fig. 8.8

SIMULTANEOUS MOTION CHART

Operation : Department : Method : Present/proposed					Date : Analyst : Time unit : Wink			
Left hand Description	**Therblig**	**Time**	**Colour**	**Clock reading**	**Colour**	**Time**	**Therblig**	**Right hand Description**

Fig. 8.9

9

Work Measurement

9.1. INTRODUCTION

Work measurement attempts to find out "How long it ought to take to complete a given job". The American Engineer F.W. Taylor is the Pioneer in the development of time study. Time study is a work measurement technique for recording the times of performing a job or its elements carried out, and for analysing the data so as to obtain the line necessary for an operator to carry it out at a defined rate of performance. It is not confined to set time standards alone by the use of stop watch. It eliminates waste of time, efforts and material. Critical analysis and close study of various operations involved in performing any task increases output without additional resources which means increase in productivity. Frank Gilbreth and his wife, Lillian Gilbreth, Pioneers in motion classification, made a major contribution in the field of time study. They developed the methods to reduce fatigue and the concept of Relaxation Allowances. Time study is the most widely used technique of work measurement. It is best suited for short-cycle repetitive jobs.

9.2. OBJECTIVES AND USES OF TIME STUDY

The objective of time study is to determine, by direct observation, the quantity of human work in a specified task, and hence to derive the standard times for the task. It increases the production of high quality product at low cost.

The time study also used to,

- compare time required by alternative methods of doing a work.
- examine a given job critically to eliminate or reduce ineffective time.
- determine the number of machine an operator can attend.
- determine staff and equipments requirements.
- make effective use of people, plant and equipment.
- provide basis for production, planning and scheduling.
- provide information for estimating tenders and delivery schedules.
- recommend a sound incentive scheme.

9.3. TECHNIQUES OF WORK MEASUREMENT

The stop watch time study is the most widely used technique to determine standard time to perform any operation. The following are the other techniques of work measurement.

- Production or interruption study
- Work sampling
- Synthesis from standard data

- Predetermined time standards (PTS)
  (*i*) Work Factor
  (*ii*) Motion time measurement (MTM)
  (*iii*) Basic motion time study (BMT)
- Structural estimating.

Method study is concerned with reducing the work content by eliminating unnecessary movement of materials or operatives and substitute good methods for poor ones.

Work measurement is concerned with investigating, reducing and subsequently eliminating ineffective time. Work measurement provides time taken in the performance of an operation or series of operations. The ineffective time is shown up and separated from effective time. Work measurement is not only used to reveal the existence of ineffective time, but also be used to set standard, times for carrying out the work.

9.4. BASIC PROCEDURE OF TIME STUDY

The following steps are necessary for the systematic carrying out the work measurement :

- SELECT the job to be studied.

The first step in time study is to select the job for studies. Following are some possible reasons for selecting the job.

(1) To establish standard time for manufacture of a new job, not previously carried out.

(2) When a method of manufacture or material has been changed and a new standard time is required.

(3) When a complaint has received from a workers' representative about the existing tight time standards, which cannot be attained.

(4) To determine the standard time for the introduction of a sound incentive scheme.

(5) When a "bottleneck" situation appears, holding up subsequent operations.

(6) When any equipment/machine appears to be idle for an excessive time or its output is low.

(7) The cost of a job appears to be excessive.

In any of the situations mentioned above, a request for time study will be made and the work. Study department will assign the work to a work study analyst.

- RECORD all the relevant data relating to the circumstances in which the work is being done.
  The second step of time study is recording of information before commencing a time study.
- EXAMINE the detailed breakdown to ensure that the most effective methods and motions are used and unproductive and foreign elements are separated from recorded data.
- MEASUREwith a stop watch the quantity of work involved in each element of the operation.
- COMPIE the standard time by rating the operator and determining the allowances.
- DEFINE the method of operation for installation in planning and production shop.

9.5. TIME STUDY EQUIPMENTS

In order to make time study, certain equipments are essential. The time study equipments are divided into two categories :

(*i*) Time measuring equipments, and

(*ii*) Auxiliary equipments.

The time measuring equipment consists of :

(1) Stop watch (2) Marsto-chron (3) Kymograph

(4) Micro-chronometer (5) Wink counter etc.

The auxiliary equipment consists of :

(1) Study board (2) Time study forms.

In addition, the time study man requires a good pencil for recording, a tachometer for measuring the speed of the machines, a small calculator to make computations from observed data.

● The Stop Watch

Two types of watches are in general use for time study, namely the mechanical and the electronic. Mechanical watches can again be subdivided into the flyback and non-flyback types, with a third type the split-hand stop watch. Electronic stop watches may be subdivided into stand-alone and electronic study board. Electronic study board include an electronic timing device.

Mechanical type stop watches may be obtained with following three types of graduated scales :

(1) recording one minute per revolution by intervals of 1/5th of a second. The small hand on the upper side gives the accumulated reading. The small hand accumulates time upto 30 minutes.

(2) recording one minute per revolution calibrated in 1/100ths of a minute, with a small hand accumulates time upto 30 minutes (the decimal minute stop watch).

(3) recording 1/100th of an hour per revolution *i.e.* larger hand makes one revolution in 1/100th of an hour. The small hand records upto one hour in 100 divisions (the decimal hour stop watch).

Fig. 9.1 shows a flyback decimal-minute stop watch which is most widely used today. The hand of small dial makes 1/30th of a revolution for each revolution of the main scale hand, and thus makes a complete turn every 30 minutes. Every tenth division is numbered to facilitate

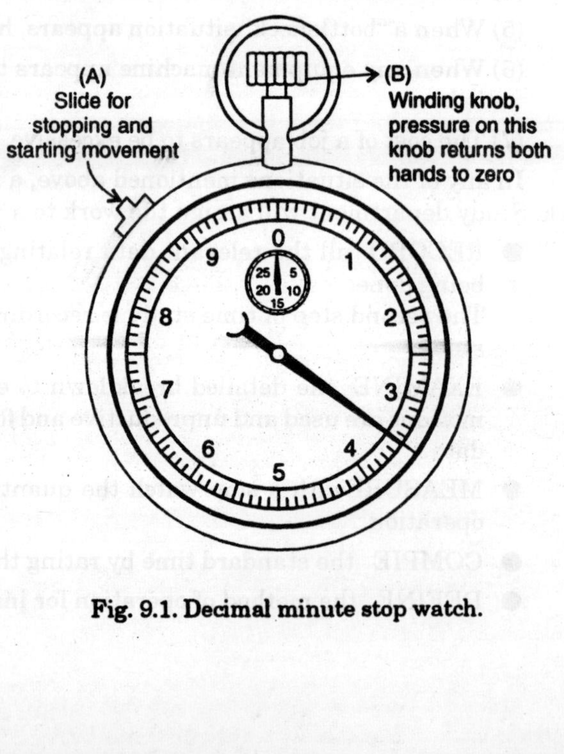

Fig. 9.1. Decimal minute stop watch.

reading. The watch is started and stopped by moving the slide '(A)' at the side of winding-knob (B). Pressure on the top of the winding-knob causes both the hands to flyback to zero. Without stopping the mechanism, from which point they immediately move forward again. If the slide (A) is used, the hands can be stopped at any point on the dial and restarted without returning to zero as soon as the slide is released. This type of watch can be used for either "flyback" or "cumulative" timing.

In non-flyback type stop watch, the first pressure on the winding knob starts the watch ; the second pressure stops it ; the third pressure returns the hands to zero. This type of watch is used for cumulative timing only.

The split hand stop watch is a more sophisticated design. In this stop watch there are two long hands moving simultaneously, one above the other. In this type of watch, pressing a secondary knob causes one of the hands to stand still while the other continuous to measure time. When the knob is pressed a second time, the stopped hand returns to the moving one and the two go on together. When a reading is taken a stopped hand is read, giving greater accuracy of reading. Split hand watch is heavier, more expensive and, due to complexity of design, it is more troublesome to repair.

● Electronic Stop Watches and Study Boards

An electronic stop watch is similar to mechanical one. It allows flyback timing to be performed without any loss of accuracy. With a mechanical watch the time taken for the hand to return to zero is "lost." With the electronic watch the timing device is running continually and it is only the display that is updated as the "fly back" button is pressed to reset the display to zero. At the end of time study it is possible to read off the total elapsed time.

Electronic study boards are used to record activity during a time study and which include an electronic timing device which automatically assigns times to elements. In this device, the observer keys a code that identifies the element being carried out and then keys in the rating assigned to the elements. At the moment of the breakpoint, observer hits a key to terminate the element. This causes the time to be recorded and the recording of the next element to begin. Time study board may have keys labelled "OC" for occasional element. "FE" for foreign element, and so on. Time is never visible to observer on screen. Important advantage to use study board is that the observer is required to carry out less writing and can pay more attention to the work being measured.

● The Study Board

The study board is usually made of wood or plastic. A study board should be light in weight and the dimensions and shape should be convenient for the work study person to hold it by the fore-arm and operate the stop watch by finger. The time study board is used to place the time study forms. It may have a fitting to hold the watch, so that the hands of the time study man are left relatively free and the watch is in a position to be read easily. The watch is normally placed at the top of the board on the right-hand side, so that the board may be rested on the left fore arm with the bottom edge against the body (Fig. 9.2). A spring clip should also be fitted to the board to hold the forms on which the study is recorded.

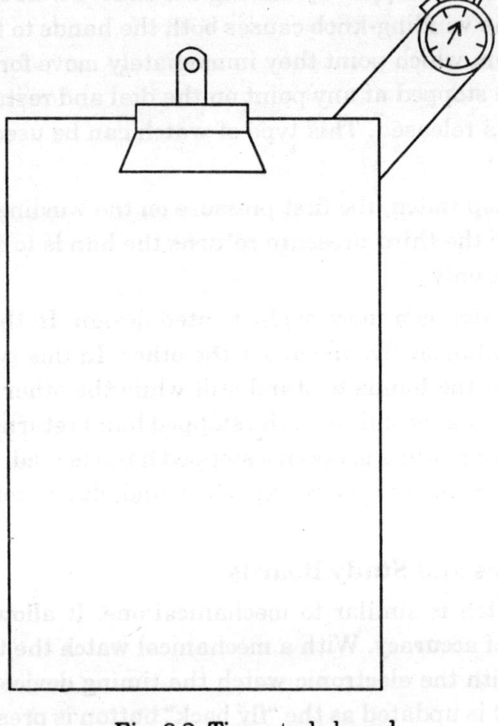

Fig. 9.2. Time study board.

- **Time Study Forms**

Time study requires the recording of substantial amounts of data. These data can be recorded on a plain paper, but it is cumbersome to rule up a new sheet every time a study is made. It is more convenient to use Pre-Printed forms which ensure that each study is made with same consistent format. There are various designs of forms. The principal forms used in time study fall into two groups ; those used at the point of observation while actually making the study, and those which are used after the study, in the study office. The forms used while actually making the study are : (*i*) Time study top sheet, (*ii*) Continuation sheet (*iii*) Short cycle study form. (Figs. 9.3, 9.4 and 9.5 on next page).

The forms used after study are : (*i*) Study summary sheet (*ii*) Analysis of studies sheet. (Figs. 9.6 and 9.7).

9.6. BREAKING AN OPERATION INTO ELEMENTS

Once the time study observer has recorded all the information about the operation and has become throughly familiar with the job he should then divide the job into small elements. An element is defined as a distinct part of a specified job selected for convenience of observation, measurement and analysis. A work cycle consists of a sequence of elements which are required to perform a job. In other words element is defined as a portion of an operation that is relatively short and has an easily identifiable 'beginning' and 'end'. A work cycle starts at the beginning of the first element of the operation and ends at the end of last element. A detailed breakdown of an operation into elements in necessary for the following reasons :

<table>
<tr><th colspan="10">Time study top sheet</th></tr>
<tr><td colspan="5" rowspan="2">Department :
Operation :
Plant/Machine :
Tools and gauges :</td><td colspan="5">Study No. :
Sheet No. : of</td></tr>
<tr><td colspan="5">Time off :
Time on :
Elapsed time :
Operative :
Clock No. :</td></tr>
<tr><td colspan="5" rowspan="2">Product/Part :
DWG No. :
Quality :</td><td colspan="5">Studied by :
Date :</td></tr>
<tr><td colspan="5">Checked :</td></tr>
<tr><td>Element Description</td><td>R</td><td>WR</td><td>ST</td><td>BT</td><td>Element Description</td><td>R</td><td>WR</td><td>ST</td><td>BT</td></tr>
<tr><td></td><td></td><td></td><td></td><td></td><td></td><td></td><td></td><td></td><td></td></tr>
<tr><td colspan="10">Note : R = Rating. WR = Watch reading. ST = Subtracted time. BT = Basic time</td></tr>
</table>

Fig. 9.3

Study No. :	Time study continuation sheet				Sheet No. : of				
Element Description	**R**	**WR**	**ST**	**BT**	**Element Description**	**R**	**WR**	**ST**	**BT**

Fig. 9.4

Short cycle study form

Department :	Section :	Study No. : Sheet No. : of
Operation : Plant/Machine : Tools and gauges :		Time off : Time on : Elapsed time : Operative : Clock No. :
Product/Part : DWG No. : Quality :	Working conditions :	Studied by : Date : Checked :

El. No.	Element Description	Observed time										Total OT	Average OT	R	BT
		1	2	3	4	5	6	7	8	9	10				

Fig. 9.5

<table>
<tr><th colspan="6">Study summary sheet</th></tr>
<tr><td colspan="4">Department :</td><td>Section :</td><td>Study No. :
Sheet No. : of</td></tr>
<tr><td colspan="5" rowspan="3">Operation :

Plant/Machine :

Tools and gauges :</td><td>Date :</td></tr>
<tr><td>Time off :
Time on :</td></tr>
<tr><td>Elapsed time :
Check time :</td></tr>
<tr><td colspan="5" rowspan="3">Product/Part :

DWG No. :

Quality :</td><td>Net time :
Obs. time :</td></tr>
<tr><td>Unacc. time
UT as %</td></tr>
<tr><td rowspan="2">Studied by :
Checked :</td></tr>
<tr><td colspan="5">Operative :</td></tr>
<tr><td>El. No.</td><td>Element Description</td><td>BT</td><td>F</td><td>Obs.</td><td></td></tr>
<tr><td></td><td></td><td></td><td></td><td></td><td></td></tr>
<tr><td colspan="6">Note : BT = Basic time. F = Frequency of occurrence per cycle. Obs. = No. of observations</td></tr>
</table>

Fig. 9.6

Analysis of studies

Operation :	Department :	Section :

Details of m/c materials etc.	
	Study No :
	Date made :
	Operative :
	Clock No. :
	Machine No. :
	Study taken by :
	No. of cycles studied :

El. No.	Element Description	Basic times	Totals	Average or selected basic time per occasion	Frequency per cycle	Basic time per cycle	Relaxation allowance Personal + fatigue	Standard time SM/SH per :
			Cycles	BM	BM	BM	%	SM

Note : El. = Element ; BM = Basic minutes ; SM = Standard minutes ; SH = Standard hours.

- To separate productive activity from unproductive activity.
- To improve the accuracy in assessment of rate of working. During a cycle of operation the operator may not work at the same pace throughout the cycle. He may be working faster and may perform some elements more quickly than others. The rating of individual elements will be more realistic than the rating of a complete cycle in such cases.
- To identify different types of element for appropriate treatment *e.g.* a foreign element can be identified and eliminated because it is useless element.
- To isolate element to give appropriate fatigue allowances consistent with the level of fatigue involved in its operation.
- To facilitate checking the method. This may become necessary it at a future date the time standard for the job is queired.
- To enable a detailed work specifications.
- To gether data for recurring elements.

9.7. TYPES OF ELEMENTS

- A repetitive element is an element which occurs in every work cycle of the job. *e.g.* the element of locating a workpiece in a holding device.
- An occasional element is an element which does not occur in every work cycle. *e.g.* adjusting the tension of a belt, receiving the instructions from supervisor. This element is a useful work and part of the job.
- Constant element is an element for which the basic time remains constant whenever it is performed.
  e.g. switch on machine, gauge diameter.
- A variable element is an element for which the basic time varies in relation to some characteristics of the product equipment or process, *e.g.* dimensions, weight, quality etc.

  Examples : Cutting a metal plate with hand hack saw (time varies with hardness and length.) ; Sweep floor (varies with area) ; push trolly of parts to next machine (varies with distance).
- A manual element is an element performed by worker manually.
- A machine element is an element automatically performed by a power driven machine. Physical or chemical process is also a machine element.
  e.g. anneal tubes, fire tiles, press car body shell to strape.
- A governing element is an element occupying a longer time that of any other element which is being performed concurrently, *e.g.* turn diameter on a lathe, while gauging from time to time.
- A foreign element is an element observed which does not form a part of the operation being studied.
  e.g. degreasing a part that has still to be machined further. It will be clear from the above definition that a repetitive element may also be a constant element, or a variable one. Similarly, a constant element may be repetitive or occasional and so on. A work study person should be conversant with these elements and should be able to distinguish them in any operation.

9.8. RULES FOR BREAKING A JOB INTO ELEMENTS

There are some rules for breaking an operation into elements. They include the following :

1. Each elements should be easily identifiable. Each element should have definite beginnings and endings so that they can be repeatedly recognized. These are known as break points. These can often be recognized by sound or vision. (*e.g.* the starting a machine, commencing a cut, putting down a tool.)

2. The duration of the element should be such that it can be conveniently recorded. Too short elements are difficult to be recorded while too long elements cannot be rated accurately. It is known from experience that the smallest element considered to be about 0.04 minutes (2.4 sec). For less highly trained observers it may be 0.07 to 0.1 min. very short elements should, if possible, be next to longer elements for accurate timings and recording. Long manual elements should be rated about every 0.33 min. (20 sec.).

3. Manual elements should be chosen so that they represent naturally unified and recognizably distinct segments of the operation. It is possible to identify the actions of reaching, grasping, moving to the work piece. The worker will probably perform all these as one natural set of motions rather than as a series of independent acts.

4. Manual elements should be separated from machine elements. Manual and machine time may be necessary to measure them separately to derive standard data.

5. Constant elements should be separated from variable elements.

The necessity for a fine break down of elements depends largely on the type of manufacturing, the nature of the job and the result desired. Breaking of an operation into elements can be easily understood by considering a simple example of drilling a ϕ 10 mm. hole in a plate. The operations can be broken down into elements as follows :

1. Pick up plate and locate in jig.
   Break point : left hand leaves jig.
2. Clamp plate in jig by clamping lever and nut.
   Break Point : Right hand leaves the clamp nut.
3. Move drill to plate
   Break point : Drill touches plate.
4. Drill ϕ 10 mm hole.
   Break point : Drill breaks through plate.
5. Raise drill from hole :
   Breakpoint : R. H. releases lever.
6. Unclamp job from jig
   Breakpoint : Right hand releases the clamp.
7. Remove plate from jig
   Breakpoint : plate dropped into the box.
8. Remove chips :
   Breakpoint : Right hand releases brush.

9.9. SAMPLE SIZE (STATISTICAL METHOD)

The number of cycles to be studied depends on the time of the cycle and the nature of the operation. There are various methods to determine the sample size such as practical approach, statistical approach. In practical approach, the number of observations based on cycle time. For short cylce (0.10 min. to 0.25 min.) the number of observations should be about 100. For medium cycle (1.0 min. to 5.0 min.) the number of observations should be around 25. For long cycle (above 45 min.) 3 or 4 observations may be enough.

To eliminate the judgement, a statistical approach has been proposed for use in the determination of the number of work cycles to be studied. The statistical approach is more accurate. In this the operator will generate a distribution of elapsed times for the element under consideration. This distribution will have a mean μ, and a standard deviation σ, and it may or may not be a normal distribution.

If repeated samples of size n are taken from such a population of elapsed times and if the average elapsed time for each sample is computed, a distribution of average elapsed times will be obtained. Assuming that the distribution will, for all practical purposes, be normal. Its mean, μ, will be equal to the mean, μ, of the distribution of individual elapsed times, and its standard deviation, $\sigma_{\bar{x}}$, will be equal to the standard deviation, σ, of the distribution of individual elapsed times divided by the square root of the sample size n, *i.e.* $\frac{\sigma}{\sqrt{n}}$.

The statistical approach is based on the belief that in this phase of the time study, the analyst wants to obtain a satisfactory estimate of the operator's average time for the element under consideration. The estimate will be made by calculating the sample average, $\overline{X}$, and using this sample average as an estimate of the population average. The sample size is equal to the number of elapsed times recorded for the element, and sample average is taken as the arithmetic mean of these elaspsed times. Now the question is that 'what must the sample size be if the analyst is to obtain a satisfactory estimate of the mean, $\overline{X}$, of the distribution of elapsed times which the operator will generate, for the work element involved ?' This sample-size requirement will dictate the number of observations that must be made.

Before the sample size is determined, the time study analyst must state what he considers a satisfactory estimate to be. He has to state (*i*) the degree of accuracy he wants to attain and (*ii*) the probability that this degree of accuracy will be attained. As an example, the analyst states that he wants the probability to be 92% that the sample average, $\overline{X}$, will be within 7% of the population average, μ. If this accuracy requirement is to be satisfied, the value of the sample average must be no less than 0.93 μ and no more than 1.07 μ. If the probability requirement is to be satisfied, the minimum acceptable value of 0.93 μ must coincide with $\mu - 1.75\ \sigma_{\bar{x}}$, and the maximum acceptable value of 1.07 must coincide with $\mu + 1.75\ \sigma_{\bar{x}}$, this is because, as can be found in area property table, percent of the sample averages in normal distribution of sample averages will have values between $\mu \pm 1.75\ \sigma_{\bar{x}}$. In

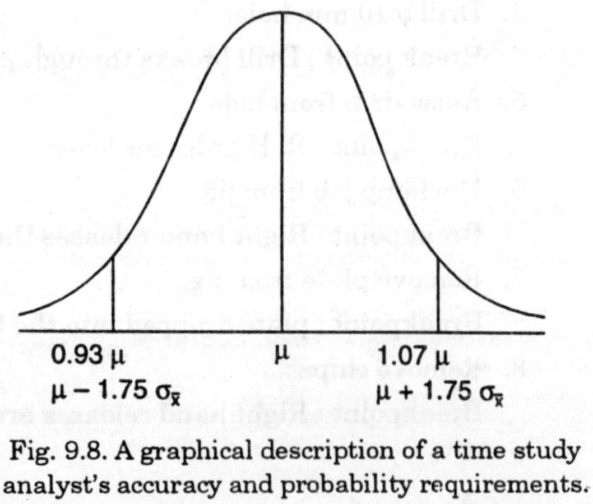

Fig. 9.8. A graphical description of a time study analyst's accuracy and probability requirements.

this case the analyst's requirement is as shown in Fig. 9.8. Algebraically, this requirement can be expressed as follows :

$$0.07\,\mu = 1.75\,\sigma_{\bar{x}} \qquad \text{...(1)}$$

This can be expressed as

$$0.07\,\mu = 1.75\left(\frac{\sigma}{\sqrt{n}}\right) \qquad \text{...(2)}$$

or

$$(0.07\,\mu)^2 = \frac{(1.75\sigma)^2}{n}$$

or

$$n = \left(\frac{1.75\sigma}{0.07\mu}\right)^2 \qquad \text{...(3)}$$

But expression (3) reveals that the analyst must know the values of the population mean and the population standard deviation before he can compute the value of the required sample size. The population mean and the population standard deviation are unknown. Therefore, analyst cannot obtain the required sample size unless he estimates their values. This is done as follows :

Here again we can apply a statistical method. For this method the analyst is required to take some arbitary number of readings. This number is an actual sample size, which is designated as 'm'. The average and the standard deviation of these 'm' elapsed times are then calculated. The estimated value of the population mean is calculated as follows :

$$\mu = \frac{\Sigma X}{m} \qquad \text{...(4)}$$

where X = observed elapsed times

m = arbitary number of readings.

$$\sigma = \sqrt{\frac{\Sigma\,(X-\mu)^2}{m}} \qquad \text{...(5)}$$

Now substituting Eq. (4) in Eq. (5), for μ,

$$\sigma = \sqrt{\frac{\Sigma\,(X-\Sigma X/m)^2}{m}} \qquad \text{...(6)}$$

Furthermore, if Eqs. (5) and (6) are substituted for μ and σ in Eq. (3), we obtain,

$$n = \left[\frac{1.75\sqrt{m\Sigma X^2 - (\Sigma X)^2}}{0.07\Sigma X}\right]^2 \qquad \text{...(7)}$$

This equation enables the analyst to obtain the sample size on the basis of observations he has already made. He can determine whether his sample size suffices to satisfy his established accuracy and probability requirements. The equation (7) is applicable for the probability requirement to be 92% and accuracy within 7% only. For the other values of probability and accuracy requirement, this can be expressed in general form as follows :

$$n = \left[\frac{Z\sqrt{m\Sigma X^2 - (\Sigma X)^2}}{D\,.\,\Sigma X}\right]^2 \qquad \text{...(8)}$$

where Z = Constant which is obtained from area property table and which reflects an established probability requirement.

D = maximum percentage deviation from the population mean which is acceptable or an established accuracy requirement.

An example will make the point clear. Let us suppose that we take five readings for a given element as given in table below. The probability requirement (confidence level) is 95% and a margin of error (accuracy requirement) applied is ± 5 percent.

Elapsed time, X in 1/100th of a min	7	6	7	7	6

We can then calculate the squares and the sum of the squares of these numbers as follows :

Sr. No.	*X*	X^2
1	7	49
2	6	36
3	7	49
4	7	49
5	6	36
$n = 5$	$\Sigma X = 33$	$\Sigma X^2 = 219$

By substituting these values in eqn. 8, we obtain the value of n.

Here probability requirement is given 95% *i.e.* 0.95. The value of Z given in area property table is for ± σ. So $\frac{0.95}{2} = 0.475$. For this value of area, we obtain the value of Z = 1.96 from the area property table. Value of D is given 5% *i.e.* 0.05,

$$\therefore \quad n = \left[\frac{1.96\sqrt{5 \times 219 - (33)^2}}{33}\right]^2$$

$$= \mathbf{8.47 \text{ or } 9 \text{ readings.}}$$

Since the number of readings 'n' that we took is less than the required sample size of nine, the sample size must be increased. However, we cannot say that four more readings are needed. When we add the values obtained from these four additional readings, the values of X, and X^2 will change, and this may affect the value of n. Consequently it may be found either that a still larger sample is required, or that the sample taken was in fact adequate or more than adequate.

9.10. PERFORMANCE RATING

Our experience shows that human abilities to do identical work differ very much ; as such the time taken to do the same job by different persons vary considerably. If this time is considered directly to find out the standard time, then there will be required separate standard time for each person working in industry. To overcome this difficulty, various methods of assessing the rate of working have been developed. The work study person will assess the rate of working of the operator by observing and relating it to standard pace. This process is known as rating. It can be defined as "the assessment of the worker's rate of working relative to the observer's concept of the rate corresponding to standard pace". It is a comparison of the rate of doing work observed by the work study person with a picture of some standard level in mind. This standard level is considered as the average rate at which the qualified workers will naturally work. This rate of working is considered as a 'standard rating', and is denoted by 100 on the rating scale. If the standard pace is maintained and the appropriate relaxation is taken, a worker will achieve 'standard performance' over the working day or shift. Standard performance is defined as the rate of output which qualified workers will naturally achieve without over-exertion as an average over the working day or shift, provided that they know and adhere to the specified method and provided that they are motivated to apply themselves to their work. This performance is denoted as 100 on the standard rating and performance scale.

A qualified worker is one who has acquired the skill, knowledge and other attributes to carry out the work in hand to satisfactory standards of quantity, quality and safety.

The rate of working as corresponding to the standard rating is equivalent to the speed of a man of average physique walking without a load in a straight line on level ground at a speed of 6.4 km/hr. Another accepted example of working at the standard rate is dealing a pack of 52 playing cards in 0.375 minutes. The standard pace will vary very widely with the environmental conditions.

9.11. SCALES OF RATING

In order to make comparison between the observed rate of working and the standard rate effectively, it is necessary to have a numerical scale against which to make the assessment. The rating is used as a factor by which the observed time can be multiplied to give the basic time. There are various scales of rating which are designated as 60-80, 75-100, 0-100 and 100-133. All the scales are linear. Which ever scale is used, the final time standards derived should be equivalent. Table 9.1 shows the various scales and corresponding rate of working.

Table 9.1. Scales of rating and corresponding activities

Scales				*Activities*	*Walking Speed km/hr*
60-80	*75-100*	*100-133*	*0-100 Standard*		
0	0	0	0	No activity	
40	50	67	50	Very slow, operator, working with no interest	3.2
60	75	100	75	Steady, unhurried performance, looks slow	4.8
80	100	133	100 Standard rating	Performance of an average, qualified worker, necessary standard & quality achieved with confidence.	6.4
100	125	167	125	Very fast, above that of an average worker	8.0
120	150	200	150	Exceptionally fast, requires intense effort.	9.6

The 0-100 scale is used in this book, in examples or discussion. In 0-100 scale, 0 represents zero activity and 100 the normal rate of working of the motivated qualified worker *i.e.* the standard rate. If the study person decides that the observed operation is being performed with less effective speed than a factor of less than 100 will be used, say 90, or 80, or 75 or whatever is considered as representing a proper assessment. If, on the other hand the study man decides that the effective rate of working is above standard, it will be given a factor greater than 100 say 110, 115, 125 or 130. It is usual practice to round off ratings to the nearest multiple of five on the scale ; that is to say, if the rate is judged to be 83 percent, it would be taken as a 85 percent.

If the study Person's rating is perfect then the result should be

$$\text{Observed time} \times \text{Rating} = \text{Constant}$$

The meaning of this can be explained by the following numerical example.

Operator	*Observed time (decimal minutes)*	*Rating %*	*Constant*
1	0.22	100	0.22
2	0.20	110	0.22
3	0.25	88	0.22

Rating is always a comparison with the standard rating (100) so that the assessed rating is the numerator of a fraction of which the denominator is the standard rating. In the above example, assessed rating is divided by standard rating 100 *i.e.* 100/100, 110/100 and so on. So that the result is obtained 0.22.

In actual practice, because of the error in judgement, rounding off to the nearest of five, the result, observed time x rating is not found constant so that this result is known as "basic time" for the element.

This can be written as

$$\text{Basic time} = \text{Observed time} \times \frac{\text{Rating}}{\text{Standard rating}}$$

$$= \text{Observed time} \times \frac{\text{Rating}}{100}$$

Basic time is the time for carrying out an element of work at standard rating. If the performance is above standard *i.e.* rating is higher than 100%, the observed time will be less than basic time. It performance is below standard, *i.e.* rating is less than 100%, the observed time will be higher than basic time.

9.12. ALLOWANCES

It is discussed in method study that the energy expended by the worker should be reduced to a minimum through the development of improved methods and procedures, in accordance with the principles of motion economy and by mechanization. Even with the improved method and use of mechanization, many jobs will still require the expenditure of human effort, and some allowance must therefore be made for recovery from fatigue and for relaxation. To make a reasonably correct estimate of the time allowed to perform an operation, unavoidable ineffective times should also be considered and added to the basic time to obtain standard time. These ineffective times may be due to waiting for materials, tool or equipment, repairs on the machine, need for rest, personnel needs etc. The first step is to eliminate or reduce this ineffective time as much as practicable. The remaining ineffective time is to be compensated by adding suitable allowances. The determination of allowances is the most controversial part of work study. It is very difficult to determine precisely the allowances needed for a given job.

The allowances can be divided into four categories as shown in Fig. 9.9. Relaxation allowances and contingency allowances are normally added to basic time. Total allowances are added into basic time to obtain standard time.

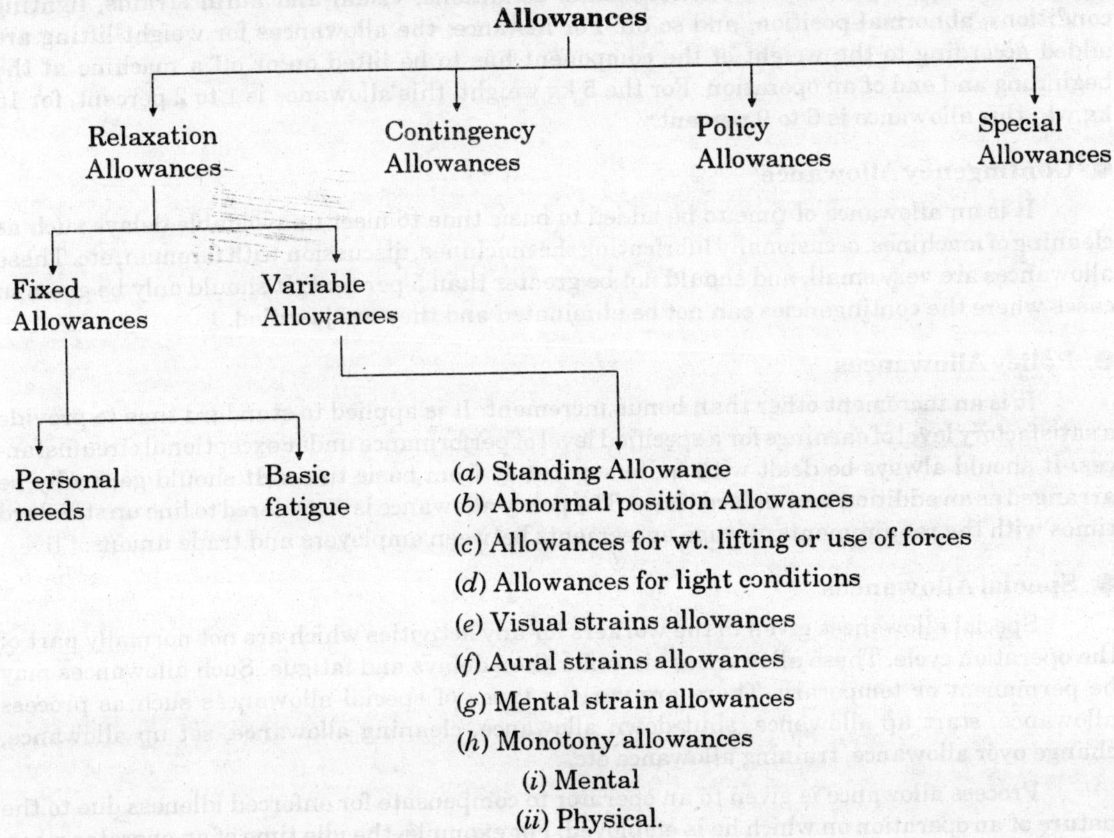

Fig. 9.9. Allowances.

● Relaxation Allowances

It is an addition to the basic time to recover the energy expended by an operator in performing a job under specified conditions. It is an allowance to recover from fatigue. Fatigue may be defined as a physical and/or mental weariness in a person and adversely affecting the ability to perform the work. Physical, chemical and physiological tests have been applied to measure fatigue. The effects of fatigue can be lessened by adding relaxation allowances to the basic time. Fatigue allowances normally added element by element so that a work value for each element is built up separately.

Relaxation allowances abbreviated as R.A. have two major components :

(i) Fixed allowances and (ii) Variable allowances.

(*i*) Fixed allowances include the personal needs of the operator such as washing, going to the lavatory etc. About 5 to 7 percent allowance is considered adequate for personal needs.

The basic fatigue allowance is also a fixed allowance. This allowance is given to take account of the energy expended while carrying out work and to alleviate monotony. The basic fatigue allowance allowed is 4 percent of the basic time is common to all operators. For the female worker this allowance is 4%.

(*ii*) Variable allowances are added to fixed allowances when working conditions differ from the normal such as poor environmental conditions, visual and aural strains, lighting conditions, abnormal position, and so on. For instance, the allowances for weight lifting are added according to the weight of the component has to be lifted on or off a machine at the beginning and end of an operation. For the 5 kg weight, this allowance is 1 to 2 percent, for 15 kg wt., this allowance is 6 to 9 percent.

● Contingency Allowance

It is an allowance of time to be added to basic time to meet unavoidable delays such as cleaning of machines, occasionally lubricating the machines, discussion with foreman, etc. These allowances are very small, and should not be greater than 5 per cent. It should only be given in cases where the contingencies can not be eliminated and they are justified.

● Policy Allowances

It is an increment other than bonus increment. It is applied to standard time to provide a satisfactory level of earnings for a specified level of performance under exceptional circumstances. It should always be dealt with quite separately from basic times. It should generally be arranged as an addition to standard times. The policy allowance is considered to line up standard times with the requirements of wage agreements between employers and trade unions.

● Special Allowances

Special allowances given to the workers for any activities which are not normally part of the operation cycle. These allowances are other than delays and fatigue. Such allowances may be permanent or temporary. There are various types of special allowances such as process allowance, start up allowance, shut down allowance, cleaning allowance, set up allowance, change over allowance, training allowance etc.

Process allowance is given to an operator to compensate for enforced idleness due to the nature of an operation on which he is employed. For example, the idle time of an operator when the machine is put on automatic cut. A start-up allowance given to compensate for time taken by any work and any enforced waiting time which occurs at the start of a shift. A shut-down allowances are similar to shut up allowances but it occurs at the end of the day. A cleaning allowance is given to worker when he has to give attention to cleaning his machine or work place. Set-up allowance, given to cover the time required for preparing a machine or process for production. Change over allowances given to compensate for time on necessary activities or waiting time at the start and/or the end of a job or batch. These allowances denoted as "job change-over allowance" or "batch change-over allowance". A training allowance given to an experienced worker to compensate for the time he is required to spend instructing a trainee.

● Standard Time

Standard time is the total time in which a job should be completed at standard performance. It is the sum of the standard times for all the element of which it is made up. It is obtained

by adding relaxation allowance and contingency allowance to the basic time. The standard time for manufacture of any component can now be computed as follows :

1. **Determine time value of each element (observed time to be recorded for each element).**
2. **Determine basic time by applying suitable rating factor.**

i.e.
$$\text{B.T.} = \frac{\text{Observed time} \times \text{Rating}}{100}$$

3. **Add allowances to basic time to obtain standard time. The standard time is expressed in standard minutes or standard hours. The standard time may be represented graphically as shown in Fig. 9.10.**

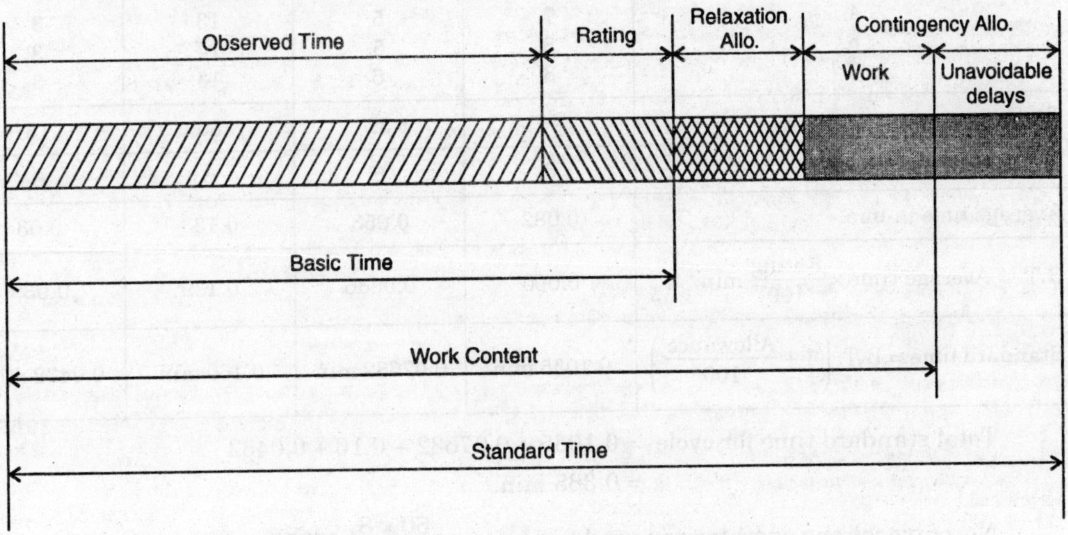

Fig. 9.10. How the standard time is made up.

Example 9.1. *The continuous time readings in centiminutes, ratings and allowances for four elements of an operation are as follows :*

Element → *Observation ↓*	*1*	*2*	*3*	*4*
1	138	43	56	59
2	67	72	85	88
3	98	204	18	21
4	28	33	46	49
5	57	62	76	79
6	87	93	306	09
Average Rating	*110%*	*120%*	*105%*	*120%*
Allowances	*15%*	*20%*	*15%*	*20%*

Calculate

(i) Standard time for each element, and for the work cycle.

(ii) Determine the number of workers required if 10,000 pieces of the above job are required per shift of 8 hours.

Solution. Following table can be prepared to obtain time for each element for all observations from the above table by subtraction.

Element →	*1*	*2*	*3*	*4*
Observation	*Subtracted time in centiminutes*			
1	8	5	13	3
2	8	5	13	3
3	10	6	14	3
4	7	5	13	3
5	8	5	14	3
6	8	6	13	3
Total time	49	32	80	18
Number of readings	6	6	6	6
Average time in min.	0.082	0.053	0.133	0.03
B.T. = Average time $\times \frac{\text{Rating}}{100}$ min.	0.090	0.0636	0.139	0.039
Standard time = B.T. $\left(1 + \frac{\text{Allowance}}{100}\right)$	0.1035 min	0.07632 min	0.160 min	0.0432 min.

$$\text{Total standard time for cycle} = 0.1035 + 0.07632 + 0.16 + 0.0432 = 0.383 \text{ min.}$$

$$\text{No. of pieces one operator can produce/day} = \frac{60 \times 8}{0.383} \approx \mathbf{1253}$$

$$\text{Staff required to produce 10,000 pieces} = \frac{10{,}000}{1253} = 7.98 \approx \mathbf{8\ persons.}$$

Example 9.2. *Time study has been carried out for a job having seven elements through cumulative timing method. From the study following observations have been made.*

Element No.	*1*	*2*	*3*	*4*	*5*	*6*	*7*
Observed time in seconds	*17*	*52*	*235*	*291*	*357*	*397*	*420*
Rating %	*105*	*115*	*90*	*100*	*85*	*120*	*80*

1. *Determine the standard time of a job considering relaxation and contingency allowances as 12% and 8% respectively of the basic time.*
2. *Estimate the observed time of the 3rd element with 105% rating and find out the rating of 5th element performed in 44 seconds.*
3. *If 25 operators work for 8 hours per day, how many days will be required to make 15,000 jobs.*

Solution.

Calculation Table

Element No.	1	2	3	4	5	6	7
Subtracted time in sec.	17	35	183	56	66	40	23
Subtracted time in min.	0.283	0.583	3.05	0.933	1.1	0.666	0.383
B.T. = Average time × $\frac{R}{100}$ min.	0.297	0.670	2.745	0.933	0.935	0.799	0.304

Total basic time = 0.297 + 0.67 + 2.745 + 0.933 + 0.935 + 0.799 + 0.304

= **6.683 minutes ≈ 401 seconds.**

Relaxation Allowance = 6.683 × 0.12 = 0.80 min.

Contingency Allowance = 6.683 × 0.08 ≈ 0.535 min.

Total Allowances = 0.80 + 0.535 = 1.335 min. ≈ 80 seconds

1. Standard time = Basic time + Allowance

= 6.683 + 1.335 = 8.018 min. ≈ 481 sec.

2. 3rd element :

$$\text{Basic time} = \text{Observed time} \times \frac{\text{Rating}}{100}$$

$$2.745 = \text{O.T.} \times \frac{105}{100}$$

$$\text{Observed time (actual)} = \frac{2.745 \times 100}{105} = 2.614 \text{ min.} \approx 157 \text{ sec.}$$

5th element :

$$\text{Basic time} = \text{Observed time} \times \frac{\text{Rating}}{100}$$

$$0.935 = 0.733 \times \frac{\text{Rating}}{100}$$

$$\therefore \quad \text{Rating} = \frac{0.935 \times 100}{0.733} = 127.55\% \approx 130\%$$

3. No. of pieces one operator can produce/day $= \frac{60 \times 8}{8.018} = 59.865 \approx 59$ pieces/day.

No. of pieces 25 operator can produce/day = 25 × 59 = 1475 pieces/day

To produce 15,000 pieces, time required $= \frac{15{,}000}{1{,}475} = 10.169 \approx$ **10 days.**

OR

Time for 1 job = 8.018 min.

Time for 15,000 job $= \frac{15000 \times 8.018}{60} =$ **2004.5 hrs.**

Time required by 25 operators $= \frac{2004.5}{25} = 80.18$ hrs.

= 10.0225 days ≈ **10 days**

Considering 8 hrs day.

10

Time Study Techniques

10.1. PRODUCTION STUDY

The stop watch time study is carried out with the qualified worker. During the study, time is recorded by study man. Environmental condition also recorded during the study and time required to perform a given operation is determined by calculation. During production, the working conditions are different and a good proportion of time is wasted due to delays and interruptions of various types. Many of these delays can be avoided by taking preventive measure. Many of these delays are unavoidable and can not be eliminated but may be reduced to a minimum by appropriate steps. Due to these unavoidable delays the operator is unable to perform the operation within specified time. To check these unavoidable delays, two methods adopted are 'production study' and 'work sampling' (ratio-delay) study.

Production study is defined as a continuous study of relatively lengthy period of one more shifts, taken with the object of checking an existing or proposed standard time, or obtaining other information. Production study is also known as interruption study. Production study is one of the techniques of work measurement. It is used to check the validity of time standards and to compare personal, fatigue, and delay allowances given during stop watch time study. In this study continuous observations of activities at an every half minute are recorded for hours together. Stop watch time study is carried out during which there is no production. Production started after obtaining the standard time by stop watch time study. Production study is carried out when production is going on in the shop. In production study, time study person will make a tour at an every half an minute in the shop and note down the activities of workers in a standard sheet. The stop watch is allowed to run continuously.

On completion of the study a summary of the productive time, preparation time and delays is prepared. The delays are also called ineffective time which is further divided into two categories—avoidable delay and unavoidable delay.

The summary also gives the time taken by the operator for personal needs. From these figures the validity of standard time issued can be checked and the fatigue can be verified. Also it is possible to verify the contingency and delay allowances.

There are many interruptions during production. Some of these interruptions are of known frequency while the others are irregular and may occur due to following reasons :

- Receiving instructions from foreman.
- Regrinding and resetting of tools.
- Cleaning and lubrication of machines.
- Periodic inspection of finished product.
- Operator going out for personal needs.
- Change of job cards after completion of a certain number of components.

It would not be possible to eliminate these interruptions but these could be timed and the frequency of their occurrence ascertained from a production study. The standard time obtain from the stop watch time study may be modified to meet the above contingencies.

10.2. WORK SAMPLING

Work sampling is defined as the method of finding the percentage occurrence of a certain activity by statistical sampling and random observations. Work sampling also known as "activity sampling", "ratio-delay study", "random observation method," "snap reading method" and "observation ratio study". Work sampling is based on the theories of sampling and probability. It is a process of drawing conclusion of large masses based on the information available from a small mass. In statistical language, the large masses is called population. A sample is a part of population. The procedure of selecting a sample, from population is called sampling.

The purpose of work sampling is to obtain a complete and accurate picture of the productive time and idle time of the machines in a specific production area. It is necessary to observe continuously all the machines in that area and to record when and why any of the machines were stopped.

In production study method continuous observations at a certain interval of time is required while work sampling is mainly based on 'probability'. Probability is designed as "the extent to which an event is likely to occur". A simple example of probability is that of tossing a coin. When we toss a coin there are two possibilities : that it will come down "heads", or that it will come down "tails". The law of probability says that there are likely to have 50% chances of head and 50% chances of tail. Infact it may be a 55-45 percent or some other ratio. But it is proved that the law becomes increasingly accurate as the number of tosses increase. This suggest that larger the size of the sample, the more accurate or representative it becomes with respect to the original "population".

Many examples can be given from our daily life where we use samples to ascertain the characteristics of the mass. Some examples are as follows :

- The purchase of a crate of mangoes from market is made by examining the upper layer of the crate. We assume that the remaining mangoes are also of the same quality.
- The purchase of a bag of grain is made by examining a sample of grain from a bag. It is expected that the sample is representative of the mass contained in the bag.
- The blood examination is made by taking few drops of blood from the body.
- A sample of rice is taken from a boiling pot and pressed between fingers to check whether the rice is properly cooked.

The sampling technique is based on the following laws :

1. Law of statistical regularity states that if a large number of items are taken at random from a group, they tend to possess the characteristics of the group.
2. Law of inertia of large numbers states that large numbers are more stable compared to small ones. For example, if the yearly production of rice is considered for Gujarat State only, the variation in output may be large due to crop failure in one district or the other ; but the total production of rice in India not vary appreciably as its

production may be more in one state even though in some other state the crop may have failed.

10.3. DETERMINATION OF SAMPLE SIZE

Statistical method

The formula used in this method is :

$$\sigma_p = \sqrt{\frac{pq}{n}} \qquad \text{...(1)}$$

where : σ_p = Standard error of proportion
p = Percentage of idle time
q = percentage of working time
n = number of observations or required sample size.

For the purpose of work sampling study, a confidence level of 95% with a 10% margin of error is considered adequate and is generally applied. This means that we are confident that in 95% of the cases our estimates will be ± 10 percent of the real value.

At the 95% confidence level, *i.e.* 0.95/2 = 0.475

From the area property table, for 0.425 area, the value is obtained 1.96.

So we can write $1.96\sigma_p = 10$

Now $\sigma_p = \sqrt{\frac{pq}{n}}$

$\therefore$ $1.96\sigma_p = 1.96\sqrt{\frac{pq}{n}}$

Take $1.96\sigma_p \approx 2\,\sigma_p$ (approx.)

$\therefore$ $2\sigma_p = 2\sqrt{\frac{pq}{n}}$

Let $2\sigma_p = \text{L} = \text{Precision required.}$

$\therefore$ $\text{L} = 2\sqrt{\frac{pq}{n}} = 2\sqrt{\frac{p(100-p)}{n}}$

because $p + q = 100\%$

$\therefore$ $(100 - p) = q$

$\therefore$ $\text{L}^2 = \frac{4p(100-p)}{n}$

$\therefore$ $n = \frac{4p(100-p)}{\text{L}^2}$

The number of observation required for a 95% confidence level can be obtained from the above formula.

The work sampling study consists of collecting data by random observations spread over the period of time. Suppose it is required to find out the idle time of a machine. In this case an observer makes a few trips in the shop and records on a tally card shown in Fig. 10.1, a particular state of machine such as idle or working.

It is observed from the tally card that total 80 observations were recorded, of which 14 times (17.5%) the machine was in idle state. Similarly 66 times (82.5%) the machine was running. If the observer in the above case had made more trips and recorded more readings, the estimate of m/c idle time would have been closer to the actual machine idle time. It is therefore necessary to decide upon the number of observations to be recorded to obtain the desired accuracy.

10.4. PROCEDURE FOR MAKING A WORK SAMPLING STUDY

The steps of work sampling study are as follows :

Tally card

Date : *Observer :*		*Study No. :*	
Number of observations : 80		*Total*	*%*
Machine running		66	82.5
Machine idle		14	17.5

Fig. 10.1. Tally Card.

(1) Classify the state of the activity into desired categories such as machine working, machine idle, tool grinding, tool setting, lubricate machine, talk with staff, talk with colleagues etc.

(2) Design a chart or observation sheet to record the above data.

(3) Make a pilot study.

(4) From the pilot study compute the delay percentage (p).

(5) Determine the accuracy and confidence level desired for study.

(6) Calculate the number of observation required for condition stated above from the equation (1).

(7) Plan the schedule of observations.

(8) Proceed with recording of observations in the chart.

(9) Analyse the data obtained.

Time required per tour of observations depends upon the number of work station, distance between two stations and the number of operators observed per station. The following numerical formula may be used to find out the time required for observation per tour.

$$t = 0.1 + 0.0164\ D + 0.04\ N$$

where t = time in min./tour

D = Distance between observation points in metres.

N = Number of operators observed per station.

10.5. ESTIMATION OF STANDARD TIME

Work sampling study is also used to estimate the standard time required to perform an operation. The standard time can be computed as follows :

1. From tally card, determine the percentage of working time.

2. From tally card, determine the percentage of idle time.
3. Find total time spent by the operator/operators on the job from time cards.
4. Estimate average performance rating of the operators.
5. Obtain the percentage of allowances given for the job under study.
6. Obtain the output during the period of study from the inspection department.

Standard time in min./component

$$= \frac{\frac{\% \text{ working time}}{100} \times \frac{\text{Average performance}}{100}}{\text{Quantity produced during observation period}} \times \left(1 + \frac{\% \text{ Allowance}}{100}\right)$$

In this technique of work measurement there is no need of any equipment. The observer makes random tours during the course of study and records the state of the subject under study, work sampling technique can be applied in the studies of warehouses, hospitals, departmental stores, work shops etc. The technique is mainly used in engineering industries to study the percentage utilization of men, machine, handling equipments in shops and to verify the allowances given for a job in stop-watch time study.

Advantages

- Not expensive as no equipment is required for gathering data.
- Trained persons are not required to collect data.
- Many activities can be studied simultaneously.
- It is less tiring as it is not required to make tour in the shop at fixed interval of time.
- The study can be interrupted any time without affecting the result.
- It is likely to be more representative as the study is spread over a number of days.
- Operations which are difficult and expensive to be measured by stop watch time study method can be measured by work sampling.

Disadvantages

Compare to advantages, the disadvantages are very few. By taking care these disadvantages may be overcome :

- It is not economical for the study of a single operation or operator.
- It is not possible to obtain elemental time data.
- The operator may change his work pattern and try to look productive, resulting in the incorrect information.
- Statistical approach of work sampling study is difficult to be understood by workers.

10.6. SYNTHESIS TECHNIQUE OF WORK MEASUREMENT

The technique used for determination of standard time from drawings and specifications of components is known as synthesis. In stop watch time study, production study and work sampling techniques, the time required to perform any operation and delays involved in the process while the job is in progress. These methods of obtaining standard time are time consuming. In addition, it is not possible to obtain the time in advance.

Synthesis is a work measurement technique for building up the time for a operation by totalling elemental times obtained previously from time studies on the other jobs containing the similar elements or from synthetic data. In this technique machining time can be obtained from graph or prepared table or it can be calculated by using the appropriate formulas. Frederick W. Taylor was the first person to divide an operation into elements, determine the time required to perform these elements and compile a dictionary of elemental times in order to facilitate estimation of time for a job. In this method, the elements are divided into two categories :

(1) Handling elements

(2) Working elements

Time of the handling elements is obtained from the previous record. Handling element are manual elements such as (*i*) pick up part and fasten in jig (*ii*) position under spindle (*iii*) release part from jig (*iv*) clean jig and table etc.

Working elements are actual machining elements. Time for machining elements is obtained from graph or may be calculated or may be obtained from standard table. Machining elements such as actual drilling time, actual turning time and so on.

Total time is obtained by adding handling time prior to machining time, machining time and handling time after machining. This total time is called basic time. The allowances are added in this basic time to obtain standard time.

Advantages of Synthesis Technique

The following are the main advantages of computing standard time for any operation from synthetic data :

- Time obtained from standard data (synthetic data) is relatively consistent because it is derived from a proven stop watch time studies and calculation.
- Time estimated by any work study man in a company will be identical as it is derived from data sheets and tables.
- Standard time for a new jobs can be computed faster as there is no need to perform the operation actually.
- It is useful for the purpose of filling tenders and quotations.
- Standard data is useful for production planning as it gives an estimate of time required to perform a job in advance.
- The use of standard data simplifies administrative work in organisations where the unions try to bargain in favour of workers.

10.7. PREDETERMINED TIME STANDARDS (PTS)

British standard glossary of terms in work study defines predetermined time standard as a work measurement technique whereby times established for basic human motions (classified according to the nature of the motion and the conditions under which it is made) are used to build up the time for a job at a defined level of performance. Predetermined time standards (PTS), also referred to as predetermined motion time system (PMTS) or synthetic time standards. As the definition indicates, PTS systems are techniques for synthesizing operations from standard time data for basic motions. In this system the time required to perform any manual operation is estimated by breaking the activity into fundamental elements and finding out the time required to perform these elements from the table. These estimates of time give basic value of time. Addition of allowances to the basic time obtained by summing all elemental time values for an operation gives the standard time for the operation.

There are six basic motions of PTS. These are as follows :

1. Reach —Move hand to destination
2. Move —Move object
3. Grasp —Secure control of object with fingers
4. Position —Line up and engaged object
5. Release —Let go of object
6. Body motions —Leg, trunk movement.

Many operations consists of some or all of these basic motions. By examining a given operation and identifying the basic motions for the operation and by referring to PTS tables for standard times for each types of motion performed under given circumstances, it is possible to derive a standard time for the operation as a whole.

The Frank Gilbreth was the pioneer of motion classification. After finding the "therbligues" by Gilbreth, the credit for adding the time dimension to motion study goes to A.B. Segur. He developed the first PTS in 1927, calling his system Motion Time Analysis.

The next important development was the work of J.H. Quick and his associates, who originated the work factor system in 1934.

A considerable number of variety of PTS systems were produced during and following the Second World War. Among these 'Methods Time Measurement', system was very widely used throughout the world.

MTM was first developed by three persons working on the system at the Westinghouse electric corporation in the U.S.A. These were H.B. Maynard, G.J. Stegemerten and J.L. Schwab. Their findings were published, and thus, for the first time, full details of a PTS system were made freely available to everyone.

10.8. WORK FACTOR SYSTEM

This system was developed by J.H. Quick, W.J. Sheer and R.E. Rochler of the work factor company Inc. work factor time data was obtained by time study analysis of motion pictures of factory operations and study of motions with stroboscopic light units. About 20,000 motions were measured by special watches for a number of years and data was recorded in a tubular form. Elemental time values given in work factor motion time tables are based on the four variables :

(1) Body member : Six body members are considered for the movements of the body, such as Arm, leg, trunk, finger or hand, foot and forearm swivel.

(2) Distance moved by body members given above. The distances are taken from a fixed point of a body member *e.g.* arm-measured at knuckle.

(3) Manual controls are classified in four categories depending upon the difficulties involved. These are : directional control (S) steer, care (P) precaution, change direction (U), and definite stop (D).

(4) Weight of resistance : The effect of weight on time depends upon the body member used and the sex of the operator. The time in general increases with increase in weight.

To obtain time value for any operation, the analyst has to break the operation into elements, identify each motion in terms of four factors given above and read the value of time from the W.F. Tables, *e.g.* move hand 25 cm. (10″) to pickup a screw driver. This element is identified as per the W.F. classification as A-10 D. From W.F. table, the time is obtained for A-10 D, 0.0042 min. The unit of time in W.F. system is one ten thousands of a minute (0.0001 min.) and the table indicates basic time. The allowances are added to the time computed from W.F. table to obtain the standard time.

10.9. METHODS TIME MEASUREMENT (MTM) SYSTEM

This system is defined as "a procedure which analyses any manual operation or method into the basic motion required to perform it and assigns to each motion a predetermined time standard which is determined by the nature of the motion and the conditions under which it is made". This system was developed from the analysis of motion, picture of the various industrial operations. The time values for fundamental motions of reach, move etc. were recorded in tabular form. The unit of time used in this method is 'Time Measurement Unit' (TMU) and is equal to 0.00001 hr. The following eight fundamental motions were considered and for each movement, time values were recorded in individual tables.

(*i*) **Reach.** Time required to reach an object depend on distance and type of reach. Reach was sub divided into five categories A to E.

(*ii*) **Move.** The time required to move depends on weight of the object and the distance moved and type of move. There are three categories of move A, B, and C.

(*iii*) & (*iv*) **Turn and Apply pressure.** The time required depends upon the degree through which the fore arm is rotated about its axis and the additional force necessary to overcome resistance.

(*v*) **Grasp.** There are five categories of grasp.

(*vi*) **Position.** The time required for position depends upon the class of fit *e.g.* loose fit, close fit, etc.

(*vii*) **Release.** There are two class of release, namely normal release and contact release.

(*viii*) **Disengage.** The time required for disengage depends upon the class of fit and ease of handling. For making a MTM study, following steps to be followed :

(1) Record all details of the job.

(2) Describe the equipment used for the manufacture of the components such as jigs, fixture etc.

(3) Break operations into elements.

(4) Identify MTM motion for each element and record them in the MTM convention such as R 250 C means reach to a component jumbled with other components and placed 250 mm away.

(5) Find time required for each element from appropriate table.

(6) Add all recorded elemental times of the operation under study. This will give basic time required for the operation.

(7) Add allowances to get the standard time.

10.10. BASIC MOTION TIME STUDY (BMT)

This system was developed by Ralph Presgrave, G.R. Barley and their associates of J.B. Woods and Gorden Limited, Toronto, during 1945 to 1951.

According to this system, the basic motion is defined as a motion in which the body member at rest moves and again comes to rest. This system was developed through laboratory study and experiments and not through film analysis of actual industrial operations.

The BMT takes into consideration, the five situations for time estimate.

(1) The distance moved during activity.

(2) The visual attention required to complete the motion.

(3) The degree of precession required to perform an activity.

(4) The force needed to overcome resistance.

(5) Simultaneous motion.

The time is obtained from BMT table. The unit of time is ten thousands (0.0001) of a minute. In this system first step is to classify the operation into element. Then record in a BMT code. As an illustration,

'Reach an object placed 200 mm away and grasp.'

This element can be written in coded form as R 200 C. From the table of 'Reach', the time is obtained.

10.11. ADVANTAGES OF PTS SYSTEMS

There are number of advantages PTS system offer over stop-watch time study. Some of them are as follows :

(1) With PTS systems one time is indicated for a given motion, irrespective of where such a motion is performed.

(2) Estimation of time for manual operations can be made without actually timing the elements of activities.

(3) A PTS system, which avoids both rating and direct observation, can lead to more consistency in setting standard times.

(4) The estimation of time to perform a job can be made from the drawing. This feature is very useful in production planning, line balancing etc.

(5) Since the times for a various operations can be derived from standard-time tables, it is possible to define the standard time for a given operation even before production begins, and often while the process is still at the design stage.

(6) PTS system allow the work study person to change the layout and design of the work place and of the necessary jigs and fixtures in such a way that the optimum production time is achieved.

(7) PTS systems are not too difficult to apply and can be less time consuming than other methods.

Inspite of the above advantages, there are some limitations to this system of time estimation. The system can be used for manual contents of jobs only. This system should be used only by throughly trained staff.

11

Incentives

11.1. INTRODUCTION

Incentives may be defined as influences which induced men to work. The aim of incentive schemes is to use man power effectively towards increased production. One of the methods used for recognition of superior performance of a worker is to maintain his enthusiasm and spirit of working by introducing incentive schemes which provide the reward to the worker for the increased production by efficient and hard work.

Direct incentive schemes based on output do not necessarily follow on an application of work measurement. There are many enterprises where time studies are made but direct incentives are not employed.

It is important for the success of any incentive scheme that the workers should know as quickly as possible the bonus they have earned. It may be shown in money units, as a percentage of the standard performance, or as the average number of standard times produced per hour.

There are mainly two categories of incentive plans :

(1) Negative incentive plans

(2) Positive incentive plans.

Negative incentive plans are punitive. In these plans, if the workers fail to attain a predetermined target of output are punished by way of payments of reduced rates, termination of services etc. The negative incentive plans are not widely followed in practice. These plans psychologically not conducive to good labour management relations. Positive incentive plans provide some benefit to better workers. There are two types of positive incentives :

(1) Non financial (2) Financial.

Non-financial incentives are shorter working hours, good working conditions, medical facilities, good accommodation, good welfare etc. These types of incentives are generally given to senior executives whose salaries are high.

11.2. NEED OF INCENTIVE PLANS

One of the greatest cause of discontent among the worker is their sense of injustice. So while introducing any incentive, it is required to be very careful. The poorly administered incentive schemes may upset the workers. It is, therefore, very essential that the financial incentive schemes are logically based, well conceived and carefully administered.

The following principles should be observed while planning the scheme :

- Before implementation of any incentive scheme, it should be seen that the job has been method studied, unnecessary movements eliminated, ineffective time reduced to a minimum and a standard time is determined.

- It should be simple to understand by workers. It should be easy to operate by management and should be acceptable to workers at large. It should benefit both the employer and employee simultaneously. Any scheme which should not favour one side more than the other side. This situation is bound to bring britterness and resentment from the side which is favoured less.
- The incentive should be substantial and generally there should be no restriction on the earnings of the workers.
- Adequate precautions should be taken to check that the quality does not suffer for quantity.
- Minimum wage should be guaranteed.

There are many types of incentive plans in industries. The widely used plans are :

(1) Day work scheme,

(2) Piece rate scheme,

(3) Premium bonus scheme, and

(4) Standard time scheme.

11.3. DAY WORK INCENTIVE SCHEME

In this system, workers are paid at a fixed rate per hour, or per day basis without consideration to his output. There are three categories of this scheme.

(*i*) Straight day work,

(*ii*) Graded day work, and

(*iii*) High day rate schemes.

In straight day work scheme a worker get money on hourly basis. Depending upon the number of hours of works, he gets money. This is very simple scheme and is easily understood by worker. He don't find any difficulty in calculating his earnings. In this system there is hardly any incentives to diligent workers. The faster workers may be at loss. To encourage faster and efficient workers graded day work scheme is useful. The workers are classified as 'A' grade, 'B' grade, 'C' grade etc. according to their speed and efficiency are fixed according to these grades thus giving more satisfaction to the efficient workers. This method increases production. The disadvantage of this system is that the grading of workers and paying them according to their merit is open to abuse as there can be lapses in merit rating.

In high day rate scheme the workers are paid at higher rate upto 50% more than the rate prevailling in that locality. This type of higher rates are offered by some industries where the nature of work is precision. Management of this type industries does not want frequent turnover of the workers. Frequent turnover of the workers affect the quality of work. In this type of industries, inefficient and non-motivated workers will not be able to reach the higher performance level set and will naturally leave the job because of very low payments. The high rate of payment prevents efficient workers from leaving the job. This system is simple to understand by the workers and easy to operate. Also there is a provision of security to the workers against sudden reduction in efficiency due to illness, accidents etc.

The disadvantages of day work scheme is that it do not give appreciable incentives. As a result they fail to satisfy ambitious workers. The cost of production is increased due to lowered production efforts.

11.4. PIECE RATE INCENTIVE SCHEME

In this scheme the payment is made proportional to output. There are three subgroup of piece rate incentive scheme.

(1) Straight piece rate,

(2) Piece rate with guaranteed minimum wage, and

(3) Differential piece rate system.

In straight piece rate scheme, the payment is made per piece produced. Worker's earning for the day is equal to the rate per piece multiplied by the number of articles produced in a day. This is also simple scheme and easily understood by the workers. Wages calculations are also simple. As the wages earned are proportional to the items produced, the production is on the higher side thus bringing down the overhead costs. Close inspection is necessary to maintain the quality of the items.

The straight piece rate scheme is modified and termed as piece rate scheme with guaranteed minimum wage. In this system the operator's minimum pay per hour is fixed. Based on this the wage per day is guaranteed. For the purpose of payment, his minimum pay per day and the rate of items prepared are considered and he gets the higher of the two figures. The direct labour cost per piece produced depends upon the production by the worker. When the output of the worker is low he has to be paid his minimum wage and hence direct labour cost is much higher. With increased production, direct labour cost reduces and remains constant at higher production rate as payment is according to the number of pieces produced by the workers.

To avoid the reduced efforts a differential piece rate incentive scheme is adopted. In this scheme there are two or more rate per piece prepared. Production upto certain level is paid at a lower rate while production above that level is paid at a higher rate per piece.

11.5. PREMIUM BONUS INCENTIVE SCHEME

This scheme was introduced by Frederick A. Halsey. In this plan the time and the production was the basis for incentive payments. Day rates are guaranteed under this plan and in addition, if the worker completes the assigned task in less than the time allowed, he is entitled to get a bonus depending upon the time saved and the type of incentive scheme.

11.6. STANDARD TIME INCENTIVE SCHEME

In this type of incentive plan, every minute saved by the worker is given to him. There is no 'profit sharing', as a result the worker puts in all his skill and efforts in performing the assigned task. He has no reason to complain as he gets money for every minute he saves. The management expect effective use of equipment and man power. The effect of this on output is considerable.

The following requirements should be satisfied before implementing this type of incentive scheme :

(1) The job should be method studied.

(2) Extreme accuracy is required in determining time standard for the job.

(3) Accurate records of time of the operator should be maintained.

It is preferable to apply this scheme when the average level of production is about 80% of the expected production.

There are many categories of the standard time incentive schemes of which the following first two are more popular while third is some time used.

(*i*) 75-100 straight proportional scheme

(*ii*) 50-100 geared scheme

(*iii*) Bedaux point system of wage incentive scheme.

In 75-100 straight proportional scheme, the time required for a job 100% performance is determined and the base rate is fixed at this performance. The operator's minimum wage is 75% of the base rate up to 75% performance from where the incentive starts.

Some times 50-100 geared scheme or differential schemes are applied instead of straight proportional scheme to bring down the slop of pay curve. In this scheme also the pay at proportion is fixed. The incentives starts at 50% performance level instead of at 75% as in previous case. Up to 75% performance, the operator earns at 75% of the basic rate fixed for 100% performance and there after the earning is proportional to the performance.

In Bedaux point system foreman is also involved to obtain good result. In this scheme of incentive payment 25% of the time saved goes to the foreman while 75% of the time saved is earned by the worker. The standard time required for each job is calculated and is given a 'B' value. Standard 'B' for one hour is 16B. Thus, if the time given for a job is 60B and an operator completes it in 50 minutes he earns 10 minutes out of which 2.5 minutes will go to the foreman while 7.5 minutes will be earned by him. The extra time earned is paid at base rate.

11.7. GROUP INCENTIVES

In this incentive scheme, the payment is made as a bonus to the group of workers in relation to their output. Such incentives are given for group work, for assembly work or for a group of operators working together as a team. In this work operators of different trades have to work together to complete the assignment. The advantage of this schemes is that the less supervision and clerical work required. There is better team spirit and co-operation among the workers, and that takes less time to train workers. There are certain drawbacks in this type of incentive plans. A faster worker suffers and a slower one earns more at the cost of faster workers. The individual worker has no control over his earning. As a result of these, if the team spirit is not good, it will lead to troubles and high efficiency can not be expected.

Profit sharing is a type of group incentive plan in which employees are accepted as partners in the enterprise. They are paid a share of profit in addition to their regular payments. Profit sharing is neither based on time taken by the workers to complete a given task nor it is based on the output of the workers, as such it can be considered as an additional remuneration. The share of profit to be distributed among the employees is fixed in advance and is usually 25% to 30% of the profit of the company.

For the successful implementation of a profit sharing plan the management has to be careful regarding the determination of the basis of profits, ascertain the net profit and device a scheme agreeable to both the employer and the employees in fixing the share of profit. In addition to these factors a method of distribution of profit also has to be carefully worked out in order to reduce conflicts.

The profit sharing plan has the following advantages :

- It improves employer-employee relations and productivity is increased.
- There is better team-work and greater co-operation among workers.
- There is a stability of workers and their turnover is reduced as the incentive is given only once or twice a year and those who leave the company are not entitled for it.

Disadvantages

- There is no direct relation between the effort and the reward. Thus the efficiency of the better workers is disregarded.
- As the rewards is given only once or twice in a year, it does not have the desired impact on the workers.
- The scheme is not sound because the profit neither results from the capital invested nor from the efforts put in by workers.
- It is difficult to determine the exact amount of profit and the workers' share.

11.8. SPECIAL FORMS OF INCENTIVE PLANS

The incentive schemes discussed so far included such categories in which the superior performance of the operator was recognised by higher rates of wage payment. This practice of incentive payment can not be justified at the present time on account of extreme sub-division of work, availability of high production machines, and superior quality of work and tool materials. The present day incentive schemes are as a consequence directed towards production of superior quality of goods, less waste of materials, better attendance etc.

- **Incentive for Quality of Product**

Management fear that the worker's interest in earning his bonus may adversely affects the quality of goods produced. One of the way to maintain quality is to ensure stringency of inspection and make payments only for approved goods. There are certain occupations such as manufacturing of decorative furniture, special brassware, handicraft etc. where quality bonus can be effectively applied.

- **Bonus for Attendance**

Irregularity and absenteeism among workers cause severe handicaps in day to day working and upset production schedules. To overcome this difficulty bonus for regularity and punctuality are implemented in many factories to improve the attendance of the workers.

- **Length of Service Bonus**

This is a type of incentive given to workers in order to prevent him to change the job. In these scheme, workers having 5 or more years of service are paid at a higher rate than the newer workers on the same job.

- **Incentive for Waste Reduction**

To reduce the waste to a minimum, the waste reduction incentive is employed. This scheme is used where the cost of materials constitutes a large proportion of the total cost. Waste reduction incentives applied in the industries such as leather, paint, wood, textile and perishable goods.

This scheme provide a bonus which is directly proportional to the saving accomplished by a worker. The bonus may be given either to an individual worker or a group of workers.

In addition to these scheme, certain other incentive schemes such as accident prevention bonus, supervisor or foreman's bonus, odd shift bonus etc. are prevalent in many organisations. The success of any incentive plan will depend upon its design, aim, implementation, profit sharing and co-operation between the employer and the employees.

12

Statistical Quality Control

12.1. STATISTICAL TOOLS IN QUALITY CONTROL

The term quality control is to mean the control of quality by statistical means. The theory of statistics and probability are brought together for this purpose. Statistics is the collection, organization, analysis, interpretation and presentation of data. The body of knowledge of statistical methods is an essential tool of the modern approach to quality.

Every manufacturing company is concerned with the quality of its product. Quality is the basic customer decision.

Customer's satisfaction is derived from quality products and services. Quality has now become an essential element of modern management.

Quality is a relative term and is used with reference to the end use of product. For example, a gear used in sugarcane juice extracting machine may not possess good surface finish, tolerance and accuracy as compared with the gear used in the head stock of a lathe, still it may consider of good quality if it works satisfactorily in the juice extracting machine.

The world "quality" has multiple meanings :

(1) Quality consists of those product features which meet the needs of customers and thereby provide product satisfaction.

(2) Quality consists of freedom from deficiencies.

(3) The component said to possess good quality, it is works well in the equipment for which it is meant. Quality is thus to define as fitness for purpose.

(4) Quality is the degree to which a specified product is preferred over competing products of equivalent grade based on comparative cost by customers.

(5) Quality of a product is a measure of fulfilment of the promises made to the customers.

Statistical sampling techniques enables the inspector upon the shop floor to take controlled random samples of components which will then be checked by measurement or gauging. The quality of the whole batch of work produced will then be judged by the results of the sample, *i.e.*, if the number of rejects found in the sample is too high then the quality of the whole batch is considered rejectable ; and otherwise the whole batch is accepted. It is like taking a few grains of boiling rice and squeezing between two fingers and judging the quality of whole rice whether it is cooked or not. It is not necessary to squeeze all the grains of rice. This is where the idea of probability involved. There is obviously a probability that the quality of few grains of rice squeeze between two fingers is different from the total quality level of the rice ; either better or worse. The statistical part of the exercise is in collecting the data from the samples, in order that the total quality level can be assessed within a certain degree of probability.

The alternative to the sampling inspection is to carry out 100% inspection. In this case every component leaving the production line is checked. This would appear to be more expensive, although the implication is that the results would have more certainty. However, this has been proved to be wrong, and there is evidence to show that 100% inspection does not guarantee that all unacceptable work will be found and rejected. This is particularly true in the case of repetitive work. In certain type of work, where the part must be destroyed in order to check its quality, statistical technique must be used.

In addition to shop floor inspection, a manufacturing concerns can use sampling techniques for checking incoming goods, such as raw material, brought out parts and sub-assemblies.

12.2. STATISTICAL CONCEPT AND VARIABILITY

Statistical concept states that no two items will be perfectly identical. There is always variation exists in nature and in industrial life. Though this may be very small variation and may be difficult to identify by human beings. For example, "identical" twins vary slightly in height and weight at birth.

Cans of tomato soup vary slightly from can to can. To disregard the existence of variation, can load to incorrect decisions on major problems. Statistics helps to analyze data properly and draw conclusions, taking into account the existence of variation. The variation in the measured parts will be due to the inherent inaccuracies of the manufacturing process, and also partly due to inaccuracies of the measuring equipment and its application. We must tolerate variabilities of size within certain limits during manufacture.

Histogram and Frequency Distribution

Let us assume that we measure the diameter of the 200 pins in a sample randomly, and obtain the following results given in Table 12.1.

Table 12.1

Diameter (mm)	*Mid size*	*Frequency*
12.1 to 12.2	12.15	4
12.2 to 12.3	12.25	21
12.3 to 12.4	12.35	50
12.4 to 12.5	12.45	70
12.5 to 12.6	12.55	43
12.6 to 12.7	12.65	12

If we plot these results on a bar-diagram having the measured dimension on the horizontal ordinate, and the frequency of occurrence of each dimension on the vertical ordinate, we shall construct a histogram. This is shown in Fig. 12.1. If a smooth curve is now drawn through the highest point of the mid-ordinate of each strip a graphical representation of the variability is obtained. This graph is called a frequency distribution curve. This curve would tend to have the same general shape if the number of pieces measured was increased, or if the grouping were varied in small degree.

Suppose now that a very much larger number of pieces were measured, and the results arranged in at least twice as many groups so that the size range of the groups become smaller ; a frequency distribution curve of more closely defined shape would emerge and would reveal the variability pattern of the production process with greater certainty.

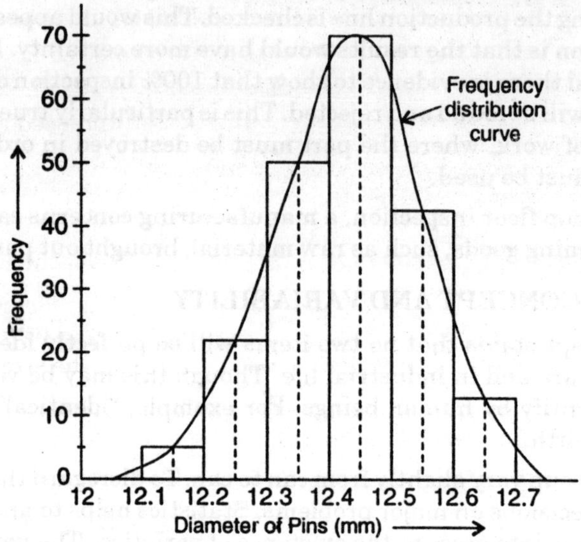

Fig. 12.1. Histogram and frequency distribution curve.

The shape of a frequency distribution curve or simply "distribution curve" is influenced by two important features of variability.

(*i*) The frequency with which parts close to the mean-size occur ;

(*ii*) The "spread" either side of the mean-size called the range.

12.3. MEAN AND SPREAD

Fig. 12.1 shows reasonably symmetrical distribution about the mean value of x. A measure of central tendency of distribution is a numerical value that describes the central position of the data or how data tend to build up in the centre. To define the distribution completely, at least two statistics are always required, (*i*) the average value or central tendency, and (*ii*) the variability or 'spread' or dispersion of the x values about the mean value. These two parameters can be calculated from the collected data. It is necessary to plot all the individual results on a graph to obtain the shape of the distribution. The most commonly used measures of 'central tendency' of the data is the arithmetic mean.

If x_1 occurs f_1 times, x_2 occurs f_2 times etc. and lastly x_n occurs f_n times there being 'n' possible different values of 'X' and 'n' observations altogether, *i.e.*

$$n = f_1 + f_2 + f_3 + \ldots\ldots + f_n$$

Then

$$\overline{X} = \frac{f_1x_1 + f_2x_2 + f_3x_3 + \ldots\ldots + f_nx_n}{f_1 + f_2 + f_3 + \ldots\ldots + f_n}$$

$$= \frac{\sum_{i=1}^{n} f_ix_i}{\Sigma f_i}$$

$$= \frac{1}{n}\Sigma f_ix_i \qquad [\because\ n = \Sigma f_i] \qquad \ldots(12.1)$$

Variability is the another measure of dispersion in SQC. This refers to the spread or scatter of x values about the mean and is most commonly measured by 'range', or the standard deviation. Range is a simple but rather crude measure of spread. The range is the difference between the largest and smallest value of a set of numbers. Expressed mathematically,

$$R = X_{max} - X_{min} \quad ...(12.2)$$

Standard Deviation (σ)

Variability is revealed by the amount of the departure from the mean size. It is roughly indicated by the range, but since range is a comparison between the size of two pieces only, it can not give an accurate measure of the variability of all pieces. The standard deviation is a more accurate measure of the variability of all the pieces because the size of every part produced is taken into account.

In measuring variability, an expression of the form,

$$f_1(x_1 - \bar{x}) + f_2(x_2 - \bar{x}) + ... + f_n(x_n - \bar{x})$$

might conceivably give what is wanted. Here $\bar{x}$ is the arithmetic mean. The positive and negative terms of this expression will be equal, and the sum of the terms zero. This difficulty is overcome by squaring the values $(x_1 - \bar{x})$ $(x_2 - \bar{x})$ ...... etc. So that there are no negative signs. The resulting value is called the 'variance'.

$$\text{Variance} = \frac{f_1(x_1 - \bar{x})^2 + f_2(x_2 - \bar{x})^2 + ... + f_n(x_n - \bar{x})^2}{N} \quad ...(12.3)$$

$$\text{Standard deviation, } \sigma = \sqrt{\text{Variance}} \quad ...(12.4)$$

Both quantities are important measures of variability. The standard deviation is of particular importance because of its relationship to the 'normal distribution' curve. The standard deviation is used when a more precise measure is desired. It is also a very common measure for quality control work and is used when subsequent statistics are to be calculated.

12.4. NORMAL DISTRIBUTION

The 200 pins considered (Fig. 12.1) will be taken from a much larger quantity manufactured. If many more pins were measured and then grouped into smaller intervals, say 0.02 mm, it would be possible to draw a histogram from which a much smoother curve of frequency distribution would result. At very large quantities and where the causes of variation are strictly random, the distribution curve will tend to be normal.

It is possible to arrive at this normal form of distribution by mathematical reasoning, the resulting equation of the normal curve is

$$F(Z) = \frac{1}{\sigma\sqrt{(2\pi)}} \cdot e^{-(x - \bar{x})^2/2\sigma^2} \quad ...(12.5)$$

where x = Size of individual part

Z = Standard normal value or frequency with which a size x-occurs

σ = Standard deviation

e = 2.71828 (constant)

$\bar{x}$ = mean value.

The equation can be simplified by putting

$$Z = \frac{(x - \bar{x})}{\sigma}$$

In order to represent the curve as simple as possible, we now put $\sigma = 1$, the equation becomes,

$$F(Z) = \frac{1}{\sqrt{(2\pi)}} \cdot e^{-Z^2/2} \qquad ...(12.6)$$

The normal curve in its standard form will have a mean of zero and standard deviation of 1. The percentage of items included between any two values can be determined by integration of equation (12.6),

i.e.

$$F(Z) = \frac{1}{\sqrt{2\pi}} \int_{-Z}^{+Z} e^{-Z^2/2}\, dZ \qquad ...(12.7)$$

The term $\frac{1}{\sqrt{2\pi}}$ included to make the area under the curve equal to unity. The area bounded by the curve and by different values of Z, expressed as a fraction of unity, then represent the probability of parts lying between these values of Z.

The term $\frac{1}{\sqrt{2\pi}}$ can be ignored in plotting a curve of normal distribution, since it only causes a change of scale on the Y-axis. The values of the area under the curve are shown in Fig. 12.2 for various values of the standard deviation.

Fig. 12.2 gives the relationship between the standard deviation and the area under the normal curve. The normal (or Gaussian) curve is symmetrical about the mean and bell shape. It is a continuous distribution like a distribution of dimensions obtained from a sample of mass produced articles. The normal distribution will therefore enable us to forecast the probability of occurrence of certain events from an engineering process.

At Fig. 12.2 it can be seen that the area under the curve subtended by the mean and one standard deviation equals 34.135%, or the area subtended by two standard deviation = 68.27%. Similarly, the area subtended by 3.92 standard deviations = 95% and by 6.18 standard deviation = 99.8%. The total area of normal distribution curve is considered 1. This means that the area under normal curve to the left of the mean is 0.5 and similarly the area to the right of the mean is 0.5. It should be noted that ordinates of ± 3.09σ do not enclose the entire area beneath the curve, because the "tail" of the curve equal to infinity. Since production patterns do not follow normal distribution exactly, the approximations that ± 3σ enclose 100% of the area, and ± 2σ enclose 95% of the area. We can say that P (any value falling between limits of ± 3σ) = 1.000, and so on for other limits. To obtain the other values of area under curve for different values of

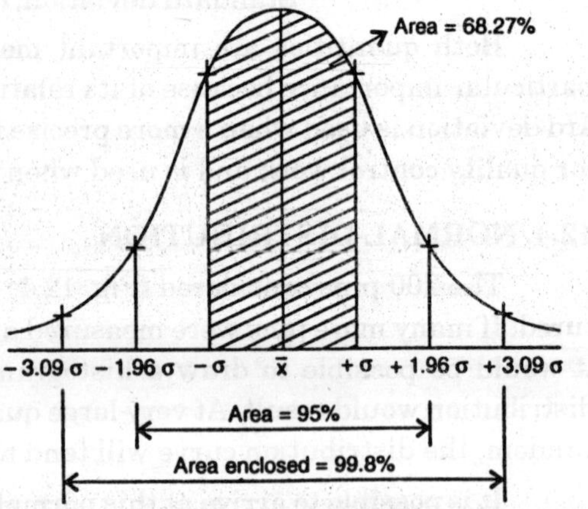

Fig. 12.2. Area properties of normal distribution curve in terms of the standard deviation (σ).

σ, the equation 12.7 can be used. However, integration is not necessary since the areas under the curve for values of Z are given in the area property table. The first step is to determine the Z value using the formula

$$Z = \frac{X_i - \overline{X}}{\sigma} \qquad \text{...(12.8)}$$

where Z = standard normal value

X_i = individual value

$\overline{X}$ = universe mean

σ = universe standard deviation.

Next, using the calculated Z value, the area under the curve to the left or right of $\bar{x}$ is found, in table of area property. (Appendix, Table 1). Let us show by means of examples how this probability distribution can be applied to measured results.

Example 12.1. *A distribution of sample results from a tensile strength test of a certain aluminium alloy is obtained, this distribution being normal having average 1780 kg/cm² and standard deviation of 225 kg/cm². What is the probability of components will have a tensile strength (i) less than 1400 kg/cm² (ii) between 1400 kg/cm² and (iii) more than 2000 kg/cm².*

Solution.

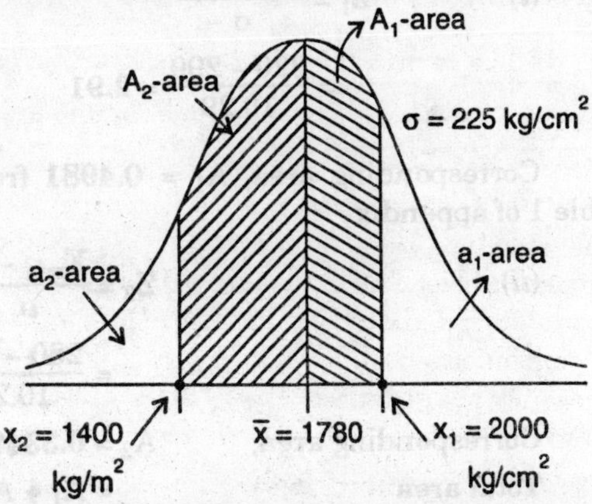

Fig. for Ex. 12.1

(*i*)
$$Z_2 = \frac{x_2 - \bar{x}}{\sigma}$$

$$= \frac{1400 - 1780}{225} \approx -1.69$$

For Z_2 = 1.69, the area A_2 of the curve is obtained from area property Table 1 (Appendix)

i.e. A_2 = **0.4545**

∴ area, a_2 = 0.5000 – 0.4545

= 0.0455 = **4.55%.**

Therefore probability of components having tensile strength less than 1400 kg/cm² will be 4.55%.

(*ii*)
$$Z_1 = \frac{x_2 - \bar{x}}{\sigma}$$

$$= \frac{2000 - 1780}{225} = 0.98$$

Area A_1 = 0.3365 from area property Table 1 (Appendix)

∴ Total area between 1400 kg/cm² and 2000 kg/cm²

= Area A_1 + Area A_2

= 0.3365 + 0.4545 = 0.7910 = **79.10%.**

Therefore probability of components having tensile strength between 1400 kg/cm² and 2000 kg/cm² will be 79.10%.

(iii) Tensile strength more than 2000 kg/cm^2

i.e. $a_1 = 0.5000 - \text{area } A_1$

$= 0.5000 - 0.3365 = 0.1635 = \mathbf{16.35\%}.$

Therefore probability of tensile strength more than 2000 kg/cm^2 will be 16.35%.

Example 12.2. *A circuit of certain electrical products contains the resistors in a series. The estimated average resistance of the circuit are 60, 80 and 150 ohms. The standard deviations of each resistors are 3, 4 and 9 ohms respectively. What per cent of resistance would meet the specifications on total resistance of 300 ± 20 ohms ?*

Solution. Mean resistance of the circuit

$= 60 + 80 + 150 = \mathbf{290\ ohms}.$

$$\sigma_{assembly} = \sqrt{\sigma_1^2 + \sigma_2^2 + \sigma_3^2}$$

$$= \sqrt{3^2 + 4^2 + 9^2} = \mathbf{10.29\ ohms}.$$

$$\left.\begin{aligned} X_{max} &= 320\ \Omega \\ X_{min} &= 280\ \text{ohms} \end{aligned}\right\} \begin{aligned} &\text{From specifications} \\ &300 \pm 20 \text{ ohms} \end{aligned}$$

(i) $$Z_1 = \frac{X_{max} - \overline{X}}{\sigma}$$

$$= \frac{320 - 290}{10.29} = \mathbf{2.91}$$

Corresponding area, $A_1 = \mathbf{0.4981}$ from Table 1 of appendix

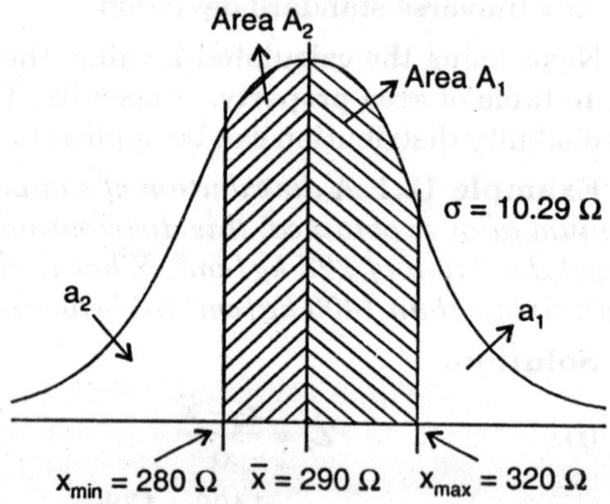

Fig. for Ex. 12.2

(ii) $$Z_2 = \frac{X_{min} - \overline{X}}{\sigma}$$

$$= \frac{280 - 290}{10.29} = \mathbf{-0.97}$$

Corresponding area, $A_2 = \mathbf{0.3340}.$

Total area $= A_1 + A_2$

$= 0.4981 + 0.3340 = \mathbf{0.8321 = 83.21\%}.$

i.e. 83.21% of resistance would meet the specification on total resistance of 300 ± 20 ohms.

12.5. CONTROL CHARTS FOR VARIABLES ($\bar{x}$ and R-Charts)

These types of control charts are used for a manufactured part which the inspector checks by measurement, not gauging. This method of control is more expensive. It reveals much more about the behaviour of the process.

The $\overline{X} - R$ charts supply a basis on which to judge the stability of the pattern of variation. To find out whether the process is in the state of control or not, control limits are set up. The control chart is usually maintained for averages and not for individual components. The distribution for individual may or may not be normal but the distribution for averages always follow normal distribution.

In any manufacturing process there is some variation from piece to piece. Two kinds of variation exists in manufacture. The variation due to chance causes and due to assignable causes. The chance causes are inevitable in any process. This inherent process variation is a characteristics of the process and is the result of random causes. These random fluctuations cause the process to deviate either side of the average. Some common causes of variations in machined metal parts are :

(*i*) Variation in the material properties,

(*ii*) Variation in the type and supply of coolent,

(*iii*) Temperature fluctuations,

(*iv*) Condition of machine,

(*v*) Faulty jigs and fixture used,

(*vi*) Careless or untrained operator,

(*vii*) Machining errors.

The assignable causes can be discovered and corrected. Two definite and common assignable causes of variability in machined metal products are :

(1) Tool wear

(2) Errors of tool setting.

To discover these causes, the results of measurements are reduced to a numerical form and a graph is plotted. These control charts are called $\overline{X}$ and R chart*s*. The $\overline{X}$ denotes the average value found or the central measurements about which the majority of the other measurements should be grouped.

12.6. CONTROL LIMITS

The control limits on the chart are so placed as to disclose the pressure or absence of the assignable causes of quality variation. This makes possible the diagnosis and correction of many production troubles and often brings substantial improvements in product quality and reduction of spoilage and rework.

Control limits are different from specification limits. They are always with reference to individual units. The control limits are ($\overline{x} \pm 3\sigma_{\overline{x}}$) where $\sigma_{\overline{x}}$ is the standard error of the average or standard deviation of the expected frequency distribution.

If plotted points on either the $\overline{x}$-chart or the R-chart fall outside the control limit, the assignable cause should be looked for, found and whenever possible eliminated.

The standard deviation of the expected frequency distribution of the averages is represented by the symbol $\sigma_{\overline{x}}$.

Regardless of the form of the universe, whether normal or otherwise, it is true that the expected $\sigma_{\overline{x}} = \dfrac{\sigma}{\sqrt{n}}$.

i.e. standard deviation of $\overline{x} = \dfrac{\sigma}{\sqrt{n}}$ ...(12.9)

Relationship between σ and $\overline{\mathbf{R}}$

The following relationship used to estimate the universe standard deviation σ.

$$\sigma = \frac{\overline{R}}{d_2} \qquad ...(12.10)$$

where σ = universe standard deviation

$\overline{R}$ = average of range

d_2 = factor obtained from Table 2 of appendix

$$\overline{\overline{x}} = \frac{\Sigma \overline{x}_i}{n} \quad ...(12.11)$$

where n = number of samples

$$\overline{R} = \frac{\Sigma R_i}{n} \quad ...(12.12)$$

$$\sigma = \frac{\overline{R}}{d_2} \quad ...(12.13)$$

Then $3\sigma_{\overline{x}}$ can be calculated from the relationship

$$\sigma_{\overline{x}} = \frac{\sigma}{\sqrt{n}}$$

$$3\sigma_{\overline{x}} = \frac{3\sigma}{\sqrt{n}}$$

$\sigma_{\overline{x}}$ = standard error of mean

$\therefore$ Upper control limit (for $\overline{x}$ chart) $= \overline{\overline{x}} + 3\sigma_{\overline{x}}$ ...(12.14)

Lower control limit (for $\overline{x}$ chart) $= \overline{\overline{x}} - 3\sigma_{\overline{x}}$ ...(12.15)

The two steps in the calculation of $3\sigma_{\overline{x}}$ might be consolidated as

$$3\sigma_{\overline{x}} = \frac{3\overline{R}}{d_2\sqrt{n}}$$

To shorten the calculation of control limits from $\overline{R}$, this factor $\frac{3}{d_2\sqrt{n}}$, the multiplier of $\overline{R}$ in the preceding calculation, has been computed for each value of n, 2 to 20 and tabulated in Table 3 of appendix and designated as A_2.

The formulas for 3-sigma control limits on charts for $\overline{x}$ then becomes,

$$UCL_{\overline{x}} = \overline{\overline{x}} + A_2\overline{R} \quad ...(12.16)$$

$$LCL_{\overline{x}} = \overline{\overline{x}} - A_2\overline{R} \quad ...(12.17)$$

12.7. CONTROL CHART FOR RANGE

The control chart for sample average is a sufficient check on the quality of the work only it is certain that the inherent process variation remains constant. It is desirable to test whether the work still being produced has the same value of σ as was used to set the limits for $\overline{x}$. R-chart shows the variability within the sample. The control limits on the chart for range (R-chart) are given by

$$UCL_R = D_4\overline{R} \quad ...(12.18)$$

$$LCL_R = D_3\overline{R} \quad ...(12.19)$$

D_3 and D_4 factors are given in Table 3 of appendix for a particular sample size.

Example 12.3. *Samples of five were taken at regular intervals from a process, 10 samples in all being taken. The results were as follows :*

Sample Number	*Dimensions in mm*				
	x_1	x_2	x_3	x_4	x_5
1	*12.63*	*12.60*	*12.62*	*12.60*	*12.65*
2	*12.62*	*12.56*	*12.61*	*12.59*	*12.60*
3	*12.60*	*12.56*	*12.59*	*12.58*	*12.63*
4	*12.61*	*12.66*	*12.63*	*12.60*	*12.61*
5	*12.59*	*12.57*	*12.58*	*12.56*	*12.59*
6	*12.59*	*12.59*	*12.61*	*12.56*	*12.57*
7	*12.59*	*12.60*	*12.60*	*12.63*	*12.57*
8	*12.61*	*12.62*	*12.60*	*12.65*	*12.60*
9	*12.58*	*12.57*	*12.60*	*12.61*	*12.60*
10	*12.54*	*12.58*	*12.62*	*12.54*	*12.56*

(i) Determine the upper control limit and lower control limit for $\overline{x}$ *and R-chart.*

(ii) Determine $\overline{\overline{x}}$ *and* $\overline{R}$ *values.*

(iii) Show the results of these samples plotted upon the chart.

Solution.

(1) Calculate $\overline{x}$ and R values for each sample and write these values in column 8 and 9 of calculation table.

$$\overline{x} = \frac{x_1 + x_2 + x_3 + x_4 + x_5}{5}$$

$$R = \text{max. value} - \text{min. value}$$

Calculation Table :

Sample No.	*Dimension in mm*					ΣX	$\overline{x} = \frac{\Sigma x}{5}$	*R*
	x_1	x_2	x_3	x_4	x_5			
1	12.63	12.60	12.62	12.60	12.65	63.1	12.62	0.05
2	12.62	12.56	12.61	12.59	12.60	62.98	12.59	0.06
3	12.60	12.56	12.59	12.58	12.63	62.96	12.59	0.07
4	12.61	12.66	12.63	12.60	12.61	63.11	12.62	0.06
5	12.59	12.57	12.58	12.56	12.59	62.89	12.57	0.03
6	12.59	12.59	12.61	12.56	12.57	62.92	12.58	0.04
7	12.59	12.60	12.60	12.63	12.57	62.99	12.59	0.06
8	12.61	12.62	12.60	12.65	12.60	63.08	12.61	0.05
9	12.58	12.57	12.60	12.61	12.60	62.96	12.59	0.04
10	12.54	12.58	12.62	12.54	12.56	62.84	12.56	0.08
						Total	125.92	0.54

(2) Add all the averages and range

i.e. $\Sigma\overline{x} = 125.92$

$\Sigma R = 0.54.$

(3) Calculate averages of averages $\bar{\bar{x}}$ and the average of range $(\bar{R})$

$$\bar{\bar{x}} = \frac{125.92}{10} = 12.592$$

$$\bar{R} = \frac{0.54}{10} = 0.054.$$

(4) For sample size of 5 from factor tables (Table 3 of appendix)

$A_2 = 0.577, \quad D_3 = 0, \quad D_4 = 2.115.$

(5) $$UCL_{\bar{x}} = \bar{\bar{x}} + A_2\bar{R}$$
$$= 12.592 + 0.577 \times 0.054$$
$$= 12.592 + 0.031158 = \mathbf{12.623}.$$

(6) $$LCL_{\bar{x}} = \bar{\bar{x}} - A_2\bar{R}$$
$$= 12.592 - 0.031158 = \mathbf{12.56}.$$

(7) $$UCL_R = D_4 . \bar{R}$$
$$= 2.115 \times 0.054 = \mathbf{0.114}.$$

(8) $$LCL_R = D_3 \times \bar{R}$$
$$= 0 \times 0.054 = 0$$

Figs. 12.3 and 12.4 show the charts.

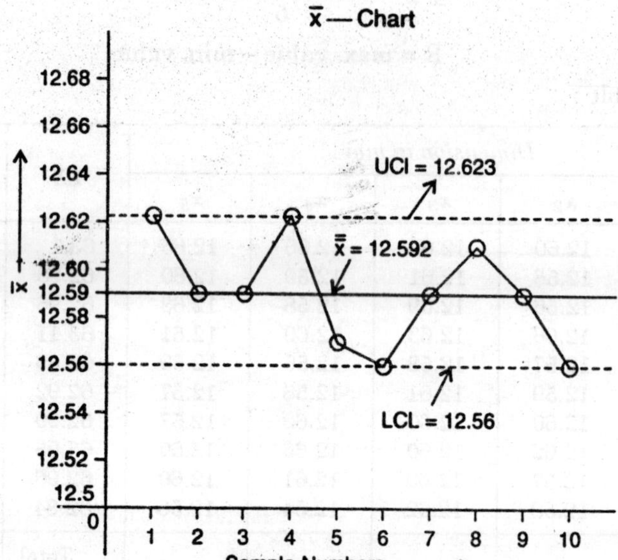

Fig. 12.3

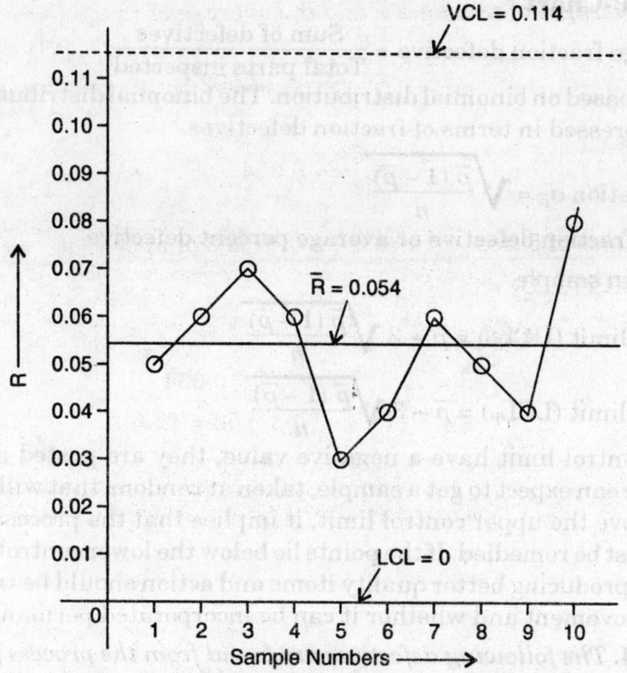

Fig. 12.4. R-chart.

12.8. CONTROL CHARTS FOR ATTRIBUTES

$\overline{X}$ and R charts are useful for the diagnosis of quality of problems and means for routine detection of source of trouble. It is used for a small fraction of quality characteristics that can be measured and expressed in numbers. $\overline{X}$ and R charts can be used only for one measurable characteristics at a time. For each dimension a separate $\overline{X}$ and R chart is necessary. It will be uneconomical and impracticable. It is preferable to analyse the results in terms of defective or non-defective terms.

Control charts for attributes are used for manufactured articles which can be classed as good or bad, *e.g.* machined parts which are being inspected with limit gauges and being classified as acceptable or defective. Sample is taken randomly. The results of samples being plotted upon the control chart. The main purpose of the chart is to indicate changes in quality so that adjustments can quickly be made to correct the process before large quantities of scrap are produced. The chart gives advance warning of the commencement of a trend towards the production of an increasing number of defective articles. There are two types of chart available for the control of attribute, these being (*i*) Fraction defective (P) chart, and (*ii*) Number defective (*n*P) chart respectively.

Whenever subgroup size is constant, the chart for fraction defective (P-chart) is used. However, if subgroup size is variable, the chart for actual number of defectives, known as '*n*P' chart is used.

Difference between defects and defectives

A defect is a quality characteristic that does not confirm to specifications.

A defective part is one that is not usable because it has one or more defects.

Control Limits for P-Chart

(1) $\bar{p}$ = average fraction defective = $\dfrac{\text{Sum of defectives}}{\text{Total parts inspected}}$ ...(12.20)

The P-chart is based on binomial distribution. The binomial distribution has the standard deviation usually expressed in terms of fraction defectives.

$$\text{Standard deviation } \sigma_{\bar{p}} = \sqrt{\frac{\bar{p}(1-\bar{p})}{n}} \quad ...(12.21)$$

where $\bar{p}$ = average fraction defective or average percent defective

n = number in sample.

$$\text{Upper control limit (UCL}_P) = \bar{p} + 3\sqrt{\frac{\bar{p}(1-\bar{p})}{n}} \quad ...(12.22)$$

$$\text{Lower control limit (LCL}_P) = \bar{p} - 3\sqrt{\frac{\bar{p}(1-\bar{p})}{n}} \quad ...(12.23)$$

If the lower control limit have a negative value, they are posted as equal to zero. The implication is that we can expect to get a sample, taken at random, that will have zero deflection. If some points lie above the upper control limit, it implies that the process is producing poorer quality items and must be remedied. If the points lie below the lower control limit the implication is that the process is producing better quality items and action should be taken to see that what has caused this improvement and whether it can be incorporated permanently in the process.

Example 12.4. *The following defectives are found from the process for 20 samples of 100 each. Plot the results in the chart :*

3, 4, 4, 3, 4, 5, 6, 10, 11, 11, 6, 6, 4, 3, 5, 8, 5, 4, 6, 5

defective respectively.

Solution.

Sr. No.	*Sample size*	*Number of defectives*	*Fraction defectives*
1	100	3	0.03
2	100	4	0.04
3	100	4	0.04
4	100	3	0.03
5	100	4	0.04
6	100	5	0.05
7	100	6	0.06
8	100	10	0.10
9	100	11	0.11
10	100	11	0.11
11	100	6	0.06
12	100	6	0.06
13	100	4	0.04
14	100	3	0.03
15	100	5	0.05
16	100	8	0.08
17	100	5	0.05
18	100	4	0.04
19	100	6	0.06
20	100	5	0.05
Total	2000	113	

$$\bar{p} = \frac{\textbf{Sum of defectives}}{\textbf{Total number inspected}}$$

$$= \frac{113}{2000} = 0.0565$$

$$\text{UCL}_p = \bar{p} + 3\sqrt{\frac{\bar{p}\,(1-\bar{p})}{n}}$$

$$= 0.0565 + 3 \times \sqrt{\frac{0.0565(1-0.0565)}{100}}$$

$$= 0.0565 + 0.06926 = \mathbf{0.1257} \text{ or } \mathbf{12.57\%}$$

$$\text{LCL}_p = 0.0565 - 0.06926$$

$$= -0.0127 \text{ or } -1.27\%$$

As LCL_p is negative, it is considered as zero.

Fig. 12.5 shows the chart.

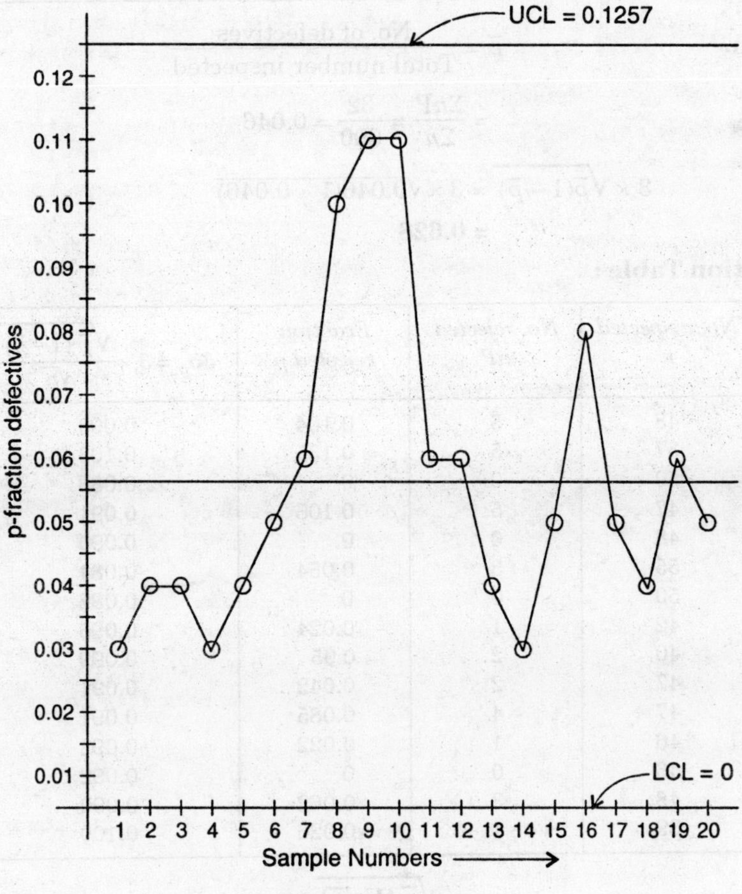

Fig. 12.5. P-chart.

12.9. NUMBER DEFECTIVE (nP chart) CHART

Example 12.5. *The following table gives data obtained from a production process for 15 samples of different sample size. Calculate the control limits for each sample and tabulate the results. Draw the distribution.*

Sample No.	*No. of units inspected*	*No. of defective units*	*Sample No.*	*No. of units inspected*	*No. of defective units*
1	48	5	9	40	2
2	37	5	10	47	2
3	50	0	11	47	4
4	47	5	12	46	1
5	48	0	13	46	0
6	55	3	14	48	3
7	50	0	15	39	1
8	42	1			
			Total	690	32

Solution.

$$\bar{p} = \frac{\text{No. of defectives}}{\text{Total number inspected}}$$

$$= \frac{\Sigma nP}{\Sigma n} = \frac{32}{690} = 0.046 \quad ...(1)$$

$$3 \times \sqrt{\bar{p}(1-\bar{p})} = 3 \times \sqrt{0.046(1-0.046)}$$

$$= \mathbf{0.628} \quad ...(2)$$

Calculation Table :

Sample No.	*No. inspected n*	*No. rejected nP*	*Fraction rejected p*	$3\sigma_p = 3 \times \frac{\sqrt{\bar{p}(1-\bar{p})}}{\sqrt{n}}$	*UCL*
1	48	5	0.104	0.090	0.136
2	37	5	0.135	0.103	0.149
3	50	0	0	0.088	0.134
4	47	5	0.106	0.091	0.137
5	48	0	0	0.090	0.136
6	55	3	0.054	0.084	0.130
7	50	0	0	0.088	0.134
8	42	1	0.024	0.096	0.142
9	40	2	0.05	0.099	0.145
10	47	2	0.042	0.091	0.137
11	47	4	0.085	0.091	0.137
12	46	1	0.022	0.092	0.138
13	46	0	0	0.092	0.138
14	48	3	0.062	0.090	0.136
15	39	1	0.026	0.100	0.146

$$3\sigma_P = 3 \times \frac{\sqrt{\bar{p}(1-\bar{p})}}{\sqrt{n}} \quad ...(3)$$

Now take '*n*' as a 48 (1st value)

$$\therefore \qquad 3\sigma_P = \frac{0.628}{\sqrt{48}} = 0.090$$

Likewise calculate the $3\sigma_p$ values for all samples and tabulate it in calculation table.

$$UCL_p = \bar{p} + 3\sigma_P$$
$$= 0.046 + 0.090 = \mathbf{0.136.}$$
$$LCL_p = \bar{p} - 3\sigma_P$$
$$= 0.046 - 0.090 = \mathbf{-0.044.}$$

As LCL is negative consider it as a zero. Calculate UCL for each sample and tabulate it. The distribution is shown in Fig. 12.6. As the sample size of variable, control limits is also variable which is not simple to interpret. To overcome this difficulties, stabilized values are calculated and tabulated.

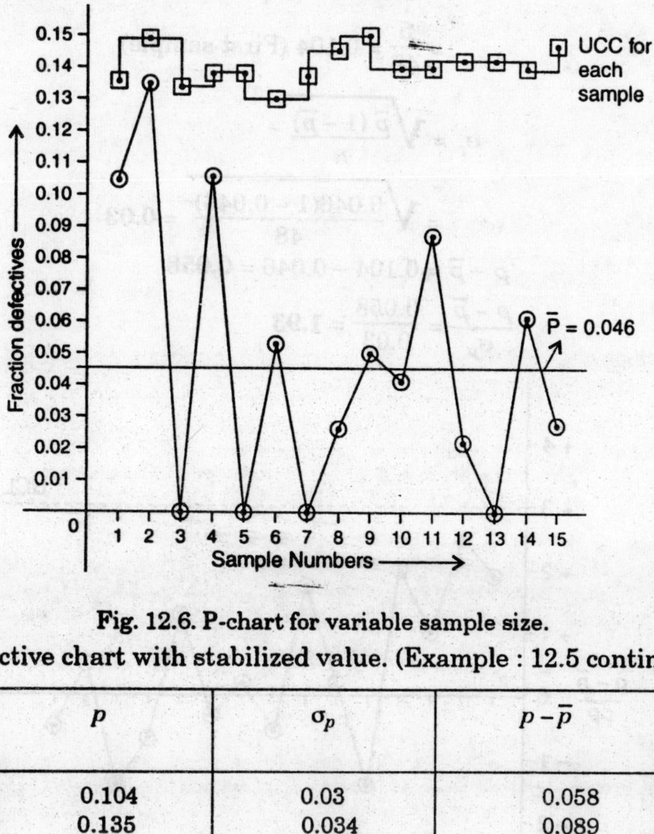

Fig. 12.6. P-chart for variable sample size.

Fraction defective chart with stabilized value. (Example : 12.5 continued)

Sample No.	p	σ_p	$p - \bar{p}$	$\frac{p - \bar{p}}{\sigma_p}$
1	0.104	0.03	0.058	1.93
2	0.135	0.034	0.089	2.62
3	0	0.029	– 0.046	1.00
4	0.106	0.030	0.06	2.00
5	0	0.030	– 0.046	– 1.53
6	0.054	0.028	0.008	0.285
7	0	0.029	– 0.046	1.58
8	0.024	0.032	– 0.022	– 0.687

Sample No.	p	σ_p	$p-\bar{p}$	$\dfrac{p-\bar{p}}{\sigma_p}$
9	0.05	0.033	0.004	0.121
10	0.042	0.030	– 0.004	– 0.133
11	0.085	0.030	0.039	1.3
12	0.022	0.031	– 0.024	– 0.774
13	0	0.031	– 0.046	– 1.48
14	0.062	0.030	0.016	0.53
15	0.026	0.033	– 0.02	– 0.608

$$\bar{p} = \frac{\Sigma nP}{\Sigma n} = 0.046$$

$$p = \frac{\text{No. defective}}{\text{Total no. inspected}}$$

$$= \frac{5}{48} = 0.104 \text{ (First sample)}$$

$$\sigma_p = \sqrt{\frac{\bar{p}(1-\bar{p})}{n}}$$

$$= \sqrt{\frac{0.046(1-0.046)}{48}} = \mathbf{0.03}$$

$$p - \bar{p} = 0.104 - 0.046 = \mathbf{0.058}$$

$$\frac{p-\bar{p}}{\sigma_p} = \frac{0.058}{0.03} = \mathbf{1.93}$$

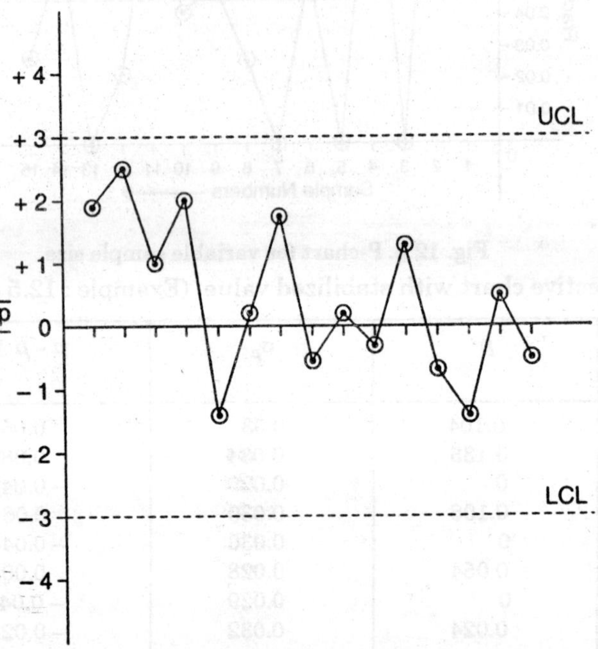

Fig. 12.7. Stabilized fraction defective chart.

Since the values are stabilized the control limits of $\pm 3\sigma_p$ can be used for comparison and it is simple to interpret.

The chart for stabilized fraction defective is shown in Fig. 12.7.

12.10. CONTROL CHART FOR DEFECTS

As defined earlier that defects means the characteristic of an item that does not meet the specification. An item is defective if it contains at least one defect. C-chart is applied to the number of defects in a sub-group of constant size. The C-chart is the another method of plotting attribute characteristics. It is a control chart that looks at the number of defects per unit of production. In most of the cases, each sub-group for C-chart consists of a single article and the variable 'C' consists of the number of defects observed in one article. However it is not necessary that the sub-group for the C-chart be a single article, it is essential that the subgroup size be constant.

Control limits for C-chart

Control limits on C-chart are based on poisson distribution. C- chart need not be restricted to a single type of defect ; but may be applicable for the total of many different kinds of defects observed on any unit.

$$\text{Average number of defects, } \overline{C} = \frac{\text{Total No. of defects in all the units}}{\text{Total No. of units}}$$

$$\text{The upper control limit, } UCL_C = \overline{C} + 3\sqrt{\overline{C}} \quad \text{...(12.25)}$$

$$\text{The lower control limit , } LCL_C = \overline{C} - 3\sqrt{\overline{C}} \quad \text{...(12.26)}$$

Example 12.6. *The data in table below gives the record of the number of defects per cabinet. Calculate the control limits for C-chart and show the result on chart.*

Cabinet No.	*Total defect in cabinet*
1	15
2	16
3	12
4	15
5	14
6	05
7	13
8	11
9	09
10	10

Total No. of defects in all the cabinet = sum of defects per cabinet = 120

Solution. $$\overline{C} = \frac{\text{Total No. of defects in all cabinets}}{\text{Total number of cabinets}}$$

$$= \frac{120}{10} = 12.0 \text{ defects/cabinet}$$

$$UCL_C = \overline{C} + 3 \times \sqrt{\overline{C}}$$

$$= 12 + 3 \times \sqrt{12} = 12 + 10.39 = \mathbf{22.39}$$

$$LCL_C = \overline{C} - 3 \times \sqrt{\overline{C}}$$
$$= 12 - 3 \times \sqrt{12} = \mathbf{1.61.}$$

The chart is shown in Fig. 12.8. No points goes outside the control limits. So process is under control.

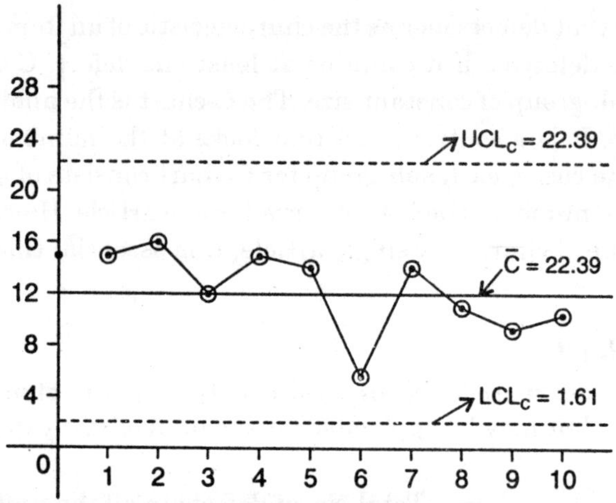

Fig. 12.8. C-chart.

12.11. ACCEPTANCE SAMPLING

In engineering or other industries, it may not be possible or practicable to inspect each one of the items. The acceptance or rejection of a lot depends on the quality levels of a small portion of the lot known as sample. 100% inspection is quite uneconomical. It involves inspection fatigue and tediousness of repetitive operations. It is not possible for destructive tests.

Acceptance sampling is usually by attributes, the parts being classed as acceptable or defective. It is economical as there are very less inspection and small number of inspectors required. Sampling plans are based on laws of probability. In acceptance sampling, the inspectors follow the prescribed sampling plan. The inspection is made without error. A sample is drawn from the lot and inspected ; if the sample is good enough then the lot is accepted, otherwise the entire lot is rejected. The most important advantage of sampling inspection is that, it exerts more effective pressure on quality improvement.

12.12. OPERATING CHARACTERISTIC CURVE

The operating characteristic (OC) curve for an attribute sampling plan is a graph of fraction defective in a lot against the probability of acceptance. A sampling plan is completely defined by its operating characteristic curve. Fig. 12.9 shows a typical OC curve with the true quality of a lot as the vertical axis and the process average defects as the horizontal axis.

Referring to Fig. 12.9 the symbols shown are :

Producer's risk (α). If the quality is good still from sampling plan some lot S are rejected and the producer has to suffer. The producer's risk is the probability of rejecting a good lot which otherwise would have been accepted. This risk is usually 0.05 or 5%. The producer can decrease his risk by producing products at a better quality level than the specified accepted quality level.

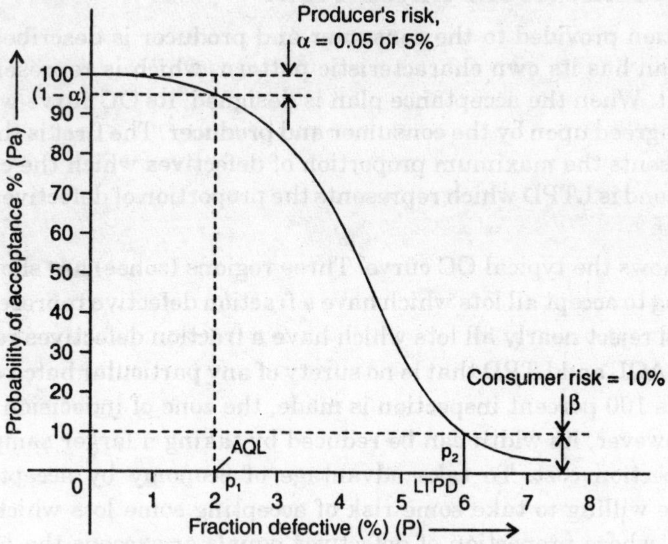

Fig. 12.9. OC curve (α = 0.05, β = 0.10).

Consumer's risk (β). If the quality is bad still there is a probability that the bad lot will be accepted by the sampling plan. In this case the consumer will suffer. Consumer's risk is the probability of defective lots being accepted which otherwise would have been rejected. Consumer's risk is usually 0.10 or 10%.

Lot Tolerance Percent Defective (LTPD) or Limiting Quality (LQ). It is also called rejectable quality level (RQL). This is a definition of unsatisfactory quality. A consumer's risk of 0.10 is common and LTPD has been defined as the lot quality for which the probability of acceptance is 0.10 of 10% of such lots will be accepted ; 90% of the lots exceeding the LTPD will be rejected. It represents the proportion of defectives which the consumer finds definitely unacceptable.

- **Acceptable Quality Level (AQL)**

It represents the maximum per cent defective or the maximum number of defects per hundred which the consumer finds definitely acceptable. A sampling plan should have a low producer's risk for quality that is equal to or better than AQL. It is the fraction defective that can be tolerated without any serious effect upon further processing. At AQL, the probability of acceptance for an AQL lot should be high.

- **Indifferent Quality Level (IQL)**

This is a quality level somewhere between the AQL and RQL. It is defined as the quality level having a probability of acceptance of 50% for a given sampling plan.

- **Average Outgoing Quality (AOQ)**

It represents the average percent defective in the outgoing products after inspection, including all accepted and all rejected lots which have been 100% inspected and defectives replaced by non-defectives.

12.13. DESIGN OF AN ACCEPTANCE PLAN

The protection provided to the consumer and producer is described by the OC curves. Each sampling plan has its own characteristic pattern, which is represented by OC curve is distinctly different. When the acceptance plan is designed, its OC curve will pass through two stipulated points agreed upon by the consumer and producer. The first is the acceptable quality level (AQL) represents the maximum proportion of defectives which the consumer will accept definitely. The second is LTPD which represents the proportion of defective which the consumer finds unacceptable.

Fig. 12.9 shows the typical OC curve. Three regions (zones) are shown in Fig. 12.9. The consumer is willing to accept all lots which have a fraction defective represented by AQL or less. The consumer will reject nearly all lots which have a fraction defectives represented by LTPD or more. Between AQL and LTPD that is no surety of any particular batch or lot being accepted or rejected. Unless 100 percent inspection is made, the zone of indecision (indifferent) cannot be eliminated. However, its width can be reduced by taking a larger sample, though this will increase the inspection cost. To take advantage of economy by acceptance sampling, the consumer must be willing to take some risk of accepting some lots which he finds definitely unacceptable, *i.e.*, whose proportion of defectives equals or exceeds the LTPD. Also producer must be willing to take some risk of having some of definitely acceptable lots rejected *i.e.* some of the lots whose proportion of defectives equals or is less than the AQL. Usually the consumer's and producer's risks are agreed upon and explicitly stated in quantitative terms. For example, the producer may agree to accept the risk that 5% of the lots with a fraction defective of say 0.02 (AQL) will be rejected, and the consumer may agree to accept the risk that 10 percent of the lots with a fraction defective of 6% (LTPD) will be accepted. (Fig. 12.9)

The curve would then be interpreted thus :

The producer's risk is 5%, that batches containing 2% defective will be rejected. The consumers risk is 10% that batches containing 6% defectives will be accepted. An O.C. curve can be plotted by finding the various Pa (probability of batches being accepted) values, for various P (fraction defectives or % defective) values.

● Characteristic of OC Curves

Figs. 12.10 and 12.11 show the comparison of OC curves for different sampling plan. In a single sampling plan three parameters are specified.

i.e. N = lot size from which the samples are drawn

n = sample size

C = acceptance number.

In Fig. 12.10 we see that with same percent samples gives very different quality protection. For example, the curves show that lots which are 4% defective lot and the lot size is 50, the probability that the lot will be accepted by the plan is 80%. However, if a lot of 200 is submitted (4% defective), then the probability of its acceptance drops to about 42%. Similarly, a lot of 1000 pieces would have only about 2% chance of passing the acceptance plan, assuming an acceptance number of zero in all cases.

Obviously a producer making a product 4% defective would have a strong motive for trying. To have his product inspected in lots of 50 rather than in lots of 100 or larger.

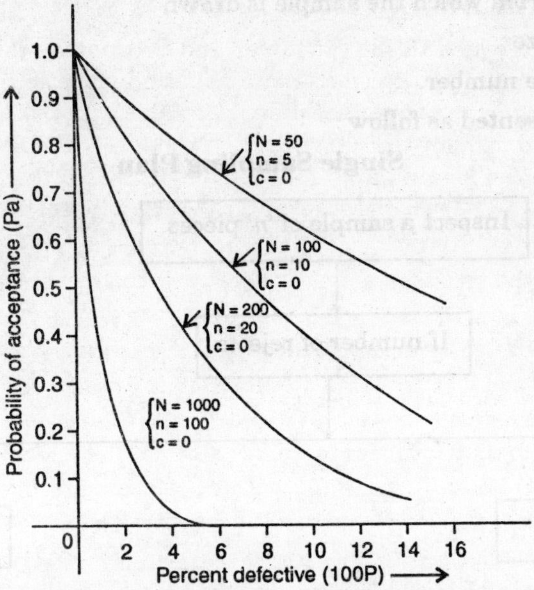

Fig. 12.10. Comparison of OC curve.

Sampling risks are affected by lot size, sample size, and the acceptance number. In Fig. 12.11 lot size is changed but the sample size and the acceptance number are kept constant. It is seen that fixed sample size tends towards constant quality protection. Variation in lot size has little effect on the probability of acceptance. If the lot size is moderately large (at least 10 times the sample size), it has small effect on the probability of acceptance. However, it is usually desired that larger lot sizes have a better operating characteristic curve in order to reduce the risk of error for large amounts of product.

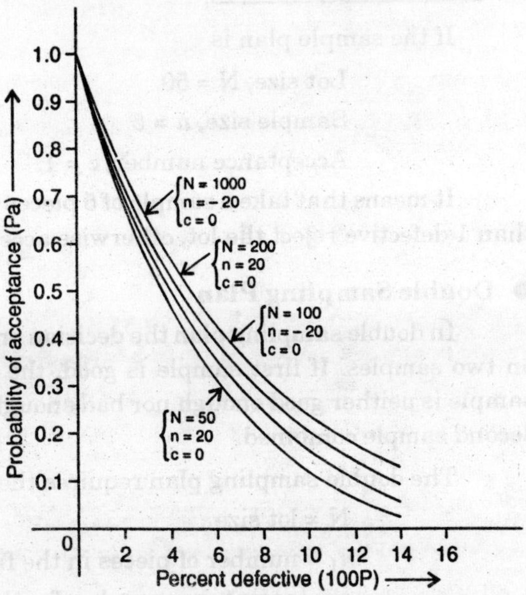

Fig. 12.11. Comparison of OC curve for four sampling plan.

Single Sampling Plan

In a single sampling plan, the decision on acceptance or rejection of the lot is based on one sample only. If the number of defectives found in the sample is equal to or less than the acceptance number, the lot or batch is accepted. If the number of defectives is equal to or greater than the rejection number, the lot or batch is rejected. In a single sampling plan, three quantities are specified.

N = lot size, from which the sample is drawn

n = sample size

c = acceptance number.

The plan is represented as follow :

Single Sampling Plan

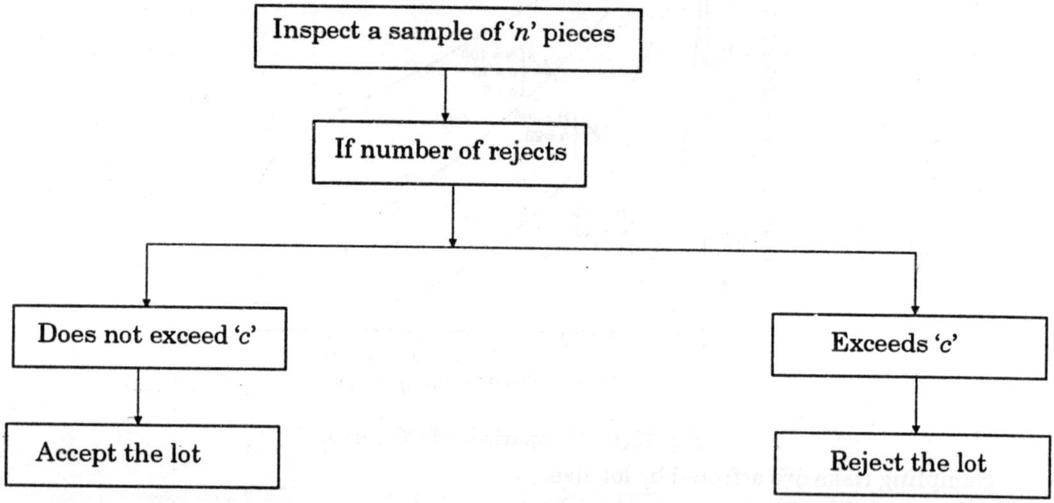

If the sample plan is

Lot size, N = 50

Sample size, $n = 5$

Acceptance number, $c = 1$.

It means that take a sample of 5 pieces from a lot of 50 pieces, if the sample contains more than 1 defective reject the lot, otherwise accept the lot.

- **Double Sampling Plan**

In double sampling plan the decision on acceptance or rejection of the lot or batch is based on two samples. If first sample is good, the lot is accepted or otherwise rejected. If the first sample is neither good enough nor bad enough the decision is taken on the evidence of first and second sample combined.

The double sampling plan requires the following parameters :

N = lot size

n_1 = number of pieces in the first sample

c_1 = acceptance number for the first sample

n_2 = number of pieces in the second sample

c_2 = acceptance number for the two samples combined

$n_1 + n_2$ = number of pieces in the two samples combined.

The plan is represented as follows :

Double Sampling Plan

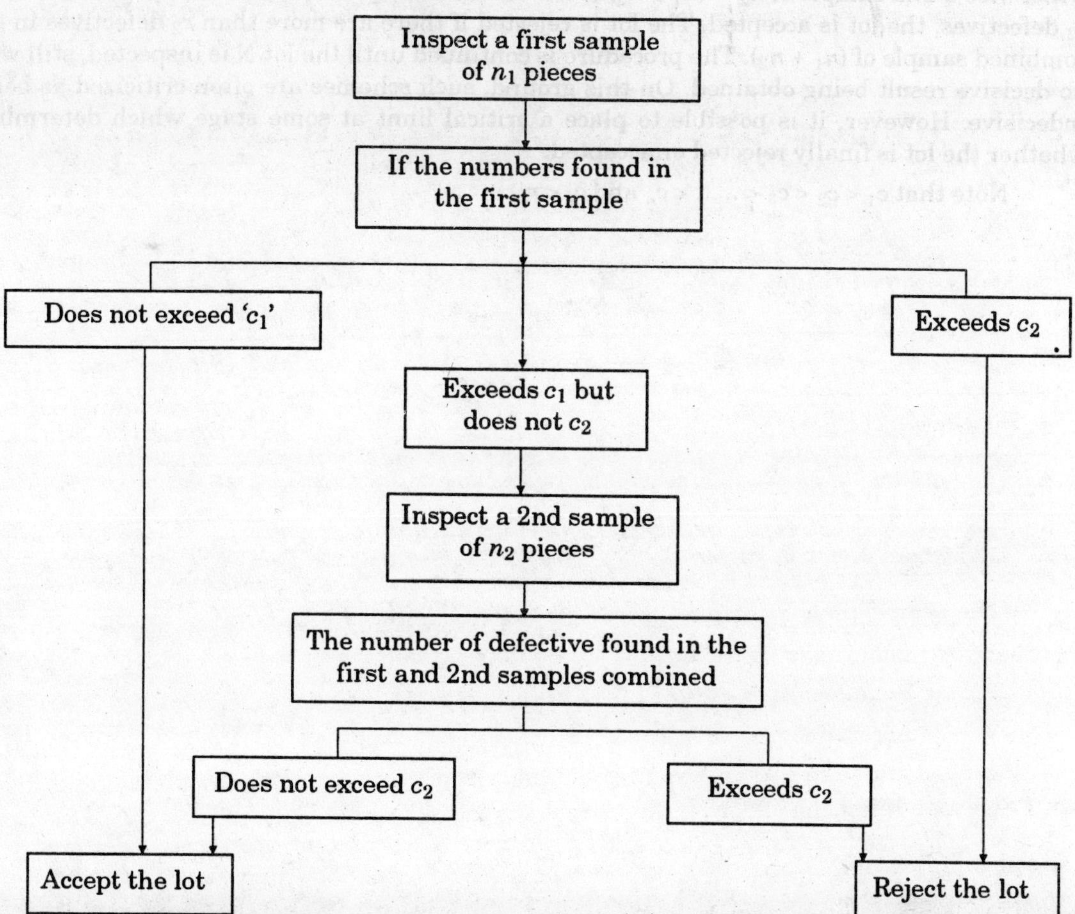

● Multiple Sampling Plan

The principle of double sampling plans can be extended to multiple sampling plans, in that many samples are taken the same value, until a decisive result for the lot is obtained. The decision on acceptance or rejection must be reached after a stated number of samples.

A multiple sampling procedure is represented in a table below :

Multiple Sampling Plan

Sample	*Sample Size*	*Combined Samples Size*	*Acceptance Number*	*Rejection Number*
1st	n_1	n_1	c_1	r_1
2nd	n_2	$n_1 + n_2$	c_2	r_2
3rd	n_3	$n_1 + n_2 + n_3$	c_3	r_3
4th	n_4	$n_1 + n_2 + n_3 + n_4$	c_4	r_4
5th	n_5	$n_1 + n_2 + n_3 + n_4 + n_5$	c_5	r_5

As shown in above table, a first sample of n_1 is drawn and inspected. If there are no more than c_1 defectives, the lot is accepted. If there are more than r_1 defectives, the lot is rejected. Otherwise a 2nd sample of $n_2 = (n_1 + n_2)$ is drawn. In this n_2 sample if there are no more than c_2 defectives, the lot is accepted. The lot is rejected if there are more than r_2 defectives in the combined sample of $(n_1 + n_2)$. The procedure is continued until the lot N is inspected, still with no decisive result being obtained. On this ground, such schemes are often criticized as being indecisive. However, it is possible to place a critical limit at some stage which determines whether the lot is finally rejected or accepted.

Note that $c_1 < c_2 < c_3 < \ldots\ldots < c_n$ and $c_i < r_i$.

Appendix

Table 1 : Area under the normal distribution curve

An entry in the table is the proportion under the entire curve which is between Z = 0 and a positive value of Z. Area for negative value of Z are obtained by symmetry.

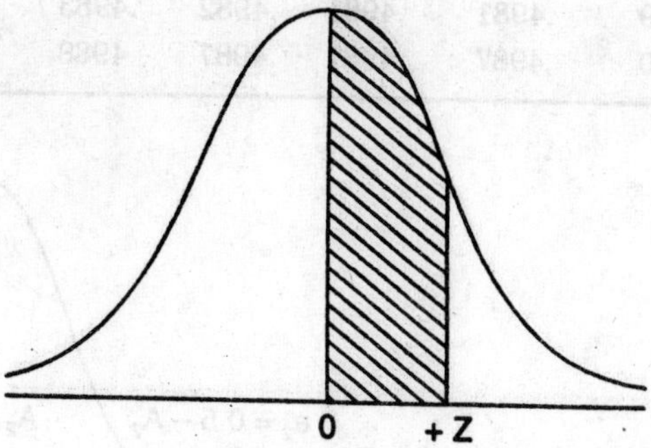

Z	.00	.01	.02	.03	.04	.05	.06	.07	.08	.09
0.0	.000	.0040	.008	.0120	.0160	.0199	.0239	.0279	.0319	.0359
0.1	.03988	.0438	.0478	.0517	.0557	.0596	.0636	.0675	.0714	.0753
0.2	.793	.0832	.0871	.0910	.0948	.0987	.1026	.1064	.1103	.1141
0.3	.1179	.1217	.1255	.1293	.1331	.1368	.1406	.1443	.1480	.1517
0.4	.1554	.1591	.1628	.1664	.1700	.1736	.1772	.1808	.1844	.1879
0.5	.1915	.1950	.1985	.2019	.2054	.2088	.2123	.2157	.2190	.2224
0.6	.2257	.2291	.2324	.2357	.2389	.2422	.2454	.2486	.2517	.2549
0.7	.2580	.2611	.2642	.2673	.2703	.2734	.2764	.2794	.2823	.2852
0.8	.2881	.2910	.2939	.2967	.2995	.3023	.3051	.3078	.3106	.3133
0.9	.3159	.3186	.3212	.3238	.3264	.3289	.3315	.3340	.3365	.3389
1.0	.3413	.3438	.3461	.3485	.3508	.3531	.3554	.3577	.3599	.3621
1.1	.3643	.3665	.3686	.3708	.3729	.3749	.3770	.3790	.3810	.3830
1.2	.3849	.3869	.3888	.3907	.3925	.3944	.3962	.3980	.3997	.4015
1.3	.4032	.4049	.4066	.4082	.4099	.4115	.4131	.4147	.4162	.4177
1.4	.4192	.4207	.4222	.4236	.4251	.4265	.4279	.4292	.4306	.4319
1.5	.4332	.4345	.4357	.4370	.4382	.4394	.4406	.4418	.4429	.4441
1.6	.4452	.4463	.4474	.4484	.4495	.4505	.4515	.4525	.4535	.4545
1.7	.4554	.4564	.4573	.4582	.4591	.4599	.4608	.4616	.4625	.4633
1.8	.4641	.4649	.4656	.4664	.4671	.4678	.4686	.4695.	.4699	.4706
1.9	.4713	.4719	.4726	.4732	.4738	.4744	.4750	.4756	.4761	.4767
2.0	.4772	.4778	.4783	.4788	.4793	.4798	.4803	.4808	.4812	.4817
2.1	.4821	.4826	.4830	.4834	.4838	.4842	.4846	.4850	.4854	.4857
2.2	.4861	.4864	.4868	.4871	.4875	.4878	.4881	.4884	.4887	.4890
2.3	.4893	.4896	.4898	.4901	.4904	.4906	.4909	.4911	.4913	.4916

2.4	.4918	.4920	.4922	.4925	.4927	.4929	.4931	.4932	.4934	.4936
2.5	.4938	.4940	.4941	.4943	.4945	.4946	.4948	.4949	.4951	.4952
2.6	.4953	.4955	.4956	.4957	.4959	.4960	.4961	.4962	.4963	.4964
2.7	.4965	.4966	.4967	.4968	.4969	.4970	.4971	.4972	.4973	.4974
2.8	.4974	.4975	.4976	.4977	.4977	.4978	.4979	.4979	.4980	.4981
2.9	.4981	.4981	.4982	.4983	.4984	.4984	.4985	.4985	.4986	.4986
3.0	.4987	.4987	.4987	.4988	.4988	.4989	.4989	.4989	.4990	.4990

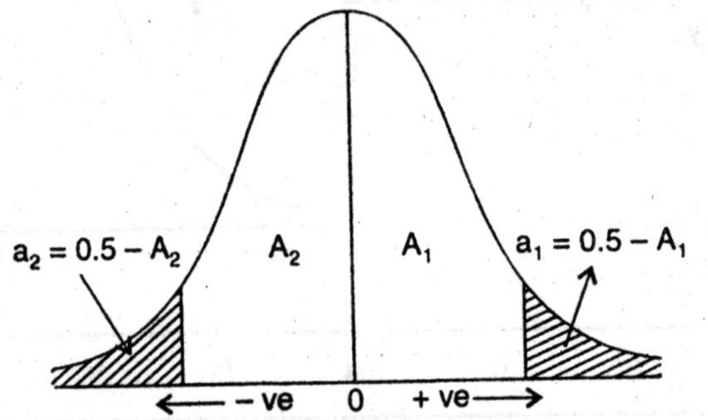

Table 2 : Factors for estimating σ from $\overline{R}$, $\overline{S}$, or σ_{RMS} and $\overline{S}$, $\overline{R}$, from $\overline{R}$

Number of observations in sub-group; n	*Factor,* $d_2 = \frac{\overline{R}}{\sigma}$	*Factor,* $d_3 = \frac{\sigma_R}{\sigma}$	*Factor* c_2, $c_2 = \frac{\sigma_{RMS}}{\sigma}$	*Factor,* $c_4 = \frac{\overline{S}}{\sigma}$
2	1.128	0.8525	0.5624	0.7979
3	1.693	0.8884	0.7236	0.8862
4	2.059	0.8798	0.7979	0.9213
5	2.326	0.8641	0.8407	0.9400
6	2.534	0.8480	0.8686	0.9515
7	2.704	0.8332	0.8882	0.9594
8	2.847	0.8198	0.9027	0.9650
9	2.970	0.8078	0.9139	0.9693
10	3.078	0.7971	0.9227	0.9727
11	3.173	0.7873	0.9300	0.9754
12	3.258	0.7785	0.9359	0.9776
13	3.336	0.7704	0.9410	0.9794
14	3.407	0.7630	0.9453	0.9810
15	3.472	0.7562	0.9490	0.9823
16	3.532	0.7499	0.9523	0.9835
17	3.588	0.7441	0.9551	0.9845
18	3.640	0.7386	0.9576	0.9854
19	3.689	0.7335	0.9599	0.9862
20	3.735	0.7287	0.9619	0.9869
21	3.778	0.7242	0.9638	0.9876
22	3.819	0.7199	0.9655	0.9882
23	3.858	0.7159	0.9670	0.9887
24	3.895	0.7121	0.9684	0.9892
25	3.931	0.7084	0.9696	0.9896
30	4.086	0.6926	0.9748	0.9914
35	4.213	0.6799	0.9784	0.9927
40	4.322	0.6692	0.9811	0.9936
45	4.415	0.6601	0.9832	0.9943
50	4.498	0.6521	0.9849	0.9949

σ_{RMS} = **Root-mean square deviation. (Std. deviation of a set of numbers)**

S = Sample standard deviation

σ = Universe standard deviation

$\overline{X}$ = Average of a sample size *n*

R = Range of sample size *n*

$\sigma_{\overline{X}}$ = Standard deviation of the averages

= Standard error of the mean

= Standard error of the average

$$\sigma = \frac{\overline{\text{R}}}{d_2} \qquad \sigma = \frac{\overline{\text{S}}}{C_4}$$

$$\text{S} = \sqrt{\frac{\Sigma(\text{X}_j - \overline{\overline{\text{X}}})^2}{n-1}}$$

$$\sigma_{\text{RMS}} = \sqrt{\frac{(\text{X}_j - \overline{\overline{\text{X}}})^2}{n}}$$

$$= \sqrt{\frac{n-1}{n}}\ .\ \text{S}.$$